DINOSAURS OF THE EAST COAST

DAVID B. WEISHAMPEL
AND LUTHER YOUNG

DINOSA

To Jack & Audrey

Dinosaurically yours,

Dave Weishampel

URS OF THE EAST COAST

THE JOHNS HOPKINS UNIVERSITY PRESS
BALTIMORE AND LONDON

June 6, 2004

© 1996 The Johns Hopkins University Press
All rights reserved. Published 1996
Printed in the United States of America on acid-free paper
9 8 7 6 5 4 3 2

The Johns Hopkins University Press
2715 North Charles Street
Baltimore, Maryland 21218-4363
www.press.jhu.edu

Library of Congress Cataloging-in-Publication Data

Weishampel, David B., 1952–
    Dinosaurs of the East Coast / David B. Weishampel and Luther
Young.
        p.   cm.
    Includes bibliographical references (p.   –   ) and index.
    ISBN 0-8018-5216-1 (alk. paper)
    1. Dinosaurs—East (U.S.)   I. Young, Luther.   II. Title.
QE862.D5W33   1996
567.9´1´0974—dc20   95-24816

A catalog record for this book is available from the British Library.

THIS BOOK IS DEDICATED TO the scientists who first brought to light the world of dinosaurs from the stone quarries, greensand pits, iron ore mines, and brickyards of the eastern states: Joseph Leidy, Edward Hitchcock, Edward Drinker Cope, Othniel Charles Marsh, John Bell Hatcher, Arthur Barneveld Bibbins, Richard Swann Lull, Mignon Talbot, and Charles Whitney Gilmore. These pioneers began the work that is still giving us our dinosaur legacy, and they inspired those who followed in this great and endlessly challenging field.

It is also dedicated to the fossil seekers of our own time, who have vigilantly saved dinosaur remains from the mining, paving, and building of the dynamic Atlantic seaboard. We owe much to these paleontologists for their renewed efforts to understand the dinosaurs of the East Coast.

CONTENTS

FOREWORD

There is a very special feeling someone gets when lucky enough to find the remains of a dinosaur. Whether it is a footprint, a single bone, or an entire skeleton, it is evidence of a spectacular animal no longer on earth.

Dinosaurs lived on every continent and all over North America. Unfortunately there are numerous places where, for some geological reason, their remains are very rare or absent. In the western United States dinosaur remains are abundant and have given us tremendous insights into the lives of these extinct creatures. In the eastern United States remains are uncommon. Only rarely do we glimpse what these Mesozoic animals looked like from more or less complete skeletons, let alone have the opportunity to understand their general behavior and lifestyles. This does not mean the eastern dinosaurs are less interesting or less important; it simply means that paleontologists working in the East have to have a great deal more patience and better luck.

As a paleontologist who has worked in the East, I can attest to the dedication of the amateur paleontological organizations that over many years have made discoveries that would help unravel the mysteries of eastern dinosaurs. This book could not have been written without their prodigious efforts.

Dinosaurs of the East Coast covers two hundred years of rare, important discoveries and interpretations. It is a book that must be read by anyone with an interest in dinosaurs or with a passion for discovery.

John R. Horner
Bozeman, Montana

PREFACE

This book has been a labor of love: love for the often-neglected world of East Coast dinosaurs, and labor to put these first-discovered American dinosaurs into their modern context.

To the degree that we have succeeded, we owe a great deal to the many dinosaur paleontologists, geologists, and dedicated fossil hunters who graciously served as field guides to the East Coast's prime fossil sites and provided us with valuable information on the dinosaurs from these localities. They include Bill Gallagher (New Jersey State Museum), Nick Fraser (Virginia Natural History Museum), Peter Dodson (University of Pennsylvania), Paul Olsen (Columbia University Lamont Doherty Earth Observatory), Hans-Dieter Sues (Royal Ontario Museum), Ed Lauginiger (Academy Park High School), Rob Weems and Ron Litwin (U.S. Geological Survey), Peter Kranz (Washington, D.C.), Jim Knight (South Carolina State Museum), Arnold Norden (Maryland Department of Natural Resources), Rich Krueger (Dinosaur State Park), Bob Denton and Bob O'Neill (New Jersey State Museum), and Mike Brett-Surman and Dan Chaney (U.S. National Museum of Natural History).

Other colleagues also generously aided our research. They include Don Baird (retired, Princeton University), Bob Sullivan (the State Museum of Pennsylvania), Vince Schneider (North Carolina State Museum), Jack

Horner (Museum of the Rockies), John Ostrom (retired, Yale University Peabody Museum), Tom Lipka (Baltimore), Neil Shubin (University of Pennsylvania), Ted Daeschler and Earl Spamer (Academy of Natural Sciences of Philadelphia), Bill Altimari (Sonora Desert Museum), Tom Holtz (U.S. Geological Survey), Ben Creisler (Seattle), Chuck Martin and Bob Purdy (U.S. National Museum of Natural History), Nick Hotton (retired, U.S. National Museum of Natural History), Dave Parris and Barbara Grandstaff (New Jersey State Museum), Dick Little (Greenfield Community College), Ken Schwartz, Jim Reger, and Emory Cleaves (Maryland Geological Survey), John Timmerman (Cape Fear Museum), Ned Colbert (Museum of Northern Arizona), Ralph Johnson (Monmouth Amateur Paleontologists Society), Donald Hoskins (Pennsylvania Topographic and Geologic Survey), Donald Ryan (Lehigh University), Walter Coombs (Western New England College), Daniel Dombroski (New Jersey Geological Survey), Richard Faas (Lafayette College), Nancy McHone (Connecticut Geological and Natural History Survey), Tom Pickett and Kelvin Ramsey (Delaware Geological Survey), Richard Foster (Massachusetts Office of Environmental Affairs), Linda Thomas (Amherst College Pratt Museum), Jeffrey Reid and Mary Watson (North Carolina Geological Survey), Duncan Heron (Duke University), Allen Penick (Virginia Division of Mineral Resources), Mike Szajna (Berks Mineral Society), Richard Dolesh (Maryland–National Capital Park and Planning Commission), Derwin Hudson (Midstate Geological Research Team), and Mary Ann Turner (Yale University Peabody Museum), who also provided us with John Bell Hatcher's field notes from his dinosaur prospecting in Maryland.

We also owe a debt of gratitude to a patient host of archivists, librarians, historians, public affairs officers, editors, reporters, artists, photographers, graduate students, property owners, business folk, and just plain friends of dinosaurs. They include Barbara Narendra and Daniel Brinkman (Yale University Peabody Museum), Lin Pennell (Philadelphia Electric Company), Murphy Smith (retired, American Philosophical Society), Tom Urquart and Jim Stimpert (the Johns Hopkins University), Charlotte Holton (American Museum of Natural History), Lawrence Lippsett (Columbia University Lamont Doherty Earth Observatory), Kathy Richards (Lehigh University), Pat Young (Science News), Daria D'Arienzo and Emily Silverman (Amherst College Archives), Don Lessem (the Dinosaur Society), Ray Ellis (Inversand Company), John Bond (Historical Consultant), Catherine Skinner (Connecticut Academy of Arts and Sciences), Thomas DeLashmutt (Oak Hill Farm), Michael Combs (Agricultural Research Service National Visitor Center), Mary Grosvald (American Association of Petroleum Geologists Library), Jack Thrasher (Above All Photo), Rosa Salter and Gene Tauber (*Allentown Morning Call*), Sydney Roby (Goucher College), Paul Payne (Maryland Clay Products), Pete Rath (retired, Maryland Clay Products), Fred Harris (Martin Marietta Aggregates), Patrick Reynolds (Historical Society of Berks County), Carol Spawn and Karen Stevens (Academy of Natural Sciences Library), Jim Bullock (government of the District of Columbia), Patricia Albright (Mount Holyoke College Archives), John Skiba (New York State Museum and Science Service), Peggy Cross (Princeton Museum of Natural History), David Wells (American Museum of Natural History), and Carlton Nash (Nash Dinosaur Tracks).

Special thanks go to our patient and thorough reviewers: Peter Dodson, Bill Gallagher, Rob Weems, Ron Litwin, Peter Kranz, Bob Sullivan, Jim Knight, Vince Schneider, and Hans-Dieter Sues.

To every one: We couldn't have done it without you.

Richard O'Grady took us up on this book project at the Johns Hopkins University Press and saw us through its early stages; Robert Harington shepherded it to fruition. We thank both for their encouragement and help.

All illustrations, unless otherwise credited, are by the authors.

Though this book has been both fun and frustrating over the course of its production, it has also required sacrifice from our families. For their encouragement, good humor, and forbearance, we gratefully thank our wives and children: Judy Weishampel, Sarah, and Amy, and Susan Yaruta Young, David, Laura, Kate, and Brooke. Now that we've got our weekends and evenings back, let's go dinosaur hunting together!

DINOSAURS OF THE EAST COAST

INTRODUCTION

To the astonishment and delight of many people, dinosaurs have been found in our own backyard in eastern North America, not just in the fertile bone beds of the American and Canadian West. Their ancient remains lie buried within the shadows of the great eastern cities. Their birdlike footprints are preserved as if these fascinating creatures walked the earth only yesterday.

Based on two centuries of fossil discovery and study, we know that dinosaurs lived along the eastern seaboard of North America throughout their 165-million-year reign during the Mesozoic era. This fossil record comes to us from many locations along the Atlantic coast: from North and South Carolina, Virginia, Washington, D.C., Maryland, Delaware, Pennsylvania, New Jersey, New York, Connecticut, Massachusetts, and Nova Scotia. (Dinosaurs that have been found elsewhere east of the Mississippi River are from the Gulf Coast province and are not included in this volume.)

The popular image of dinosaur discovery is shaped by the gigantic, exotic skeletons that are showcased in natural history museums worldwide. These exhibits appear to be the pinnacle of achievement in dinosaur paleontology. But the scientists who labor in this demanding and constantly changing field know they must look beyond such exquisite fossils and use every piece of information available to paint a fuller picture of the long-vanished past. The

1

smallest scrap of bone, a single tooth, a faint footprint, may provide critical new knowledge about dinosaurs and how they lived.

That perspective is the foundation of dinosaur paleontology on the East Coast. It is a unique story told in a unique fossil record: an intriguing combination of abundant dinosaur footprints and rare, fragmentary bones and teeth. In the two hundred years since the search began, only a handful of partial dinosaur skeletons have been found in the East, and not a trace of these animals is known from the last two-thirds of the Jurassic Period, some 180 to 140 million years ago (Ma). West Virginia, Rhode Island, Maine, Vermont, and New Hampshire have yielded no dinosaur fossils at all.

Yet the East Coast dinosaur fossil record is rich and significant, both historically and scientifically. North America's first dinosaur footprints and first dinosaur bones were discovered in the East in the early 1800s. The first nearly complete dinosaur skeleton in the world was found in New Jersey and described in 1858. The Maryland–Washington, D.C., area boasts the only Early Cretaceous dinosaur remains east of the Mississippi River, and the Connecticut Valley is one of the richest sources of dinosaur footprints in the world.

The East Coast was the cradle of North American dinosaur paleontology. It was through the efforts of such early eastern paleontologists as Joseph Leidy, Edward D. Cope, Othniel C. Marsh, Edward B. Hitchcock, and Richard Swann Lull that a solid foundation was laid for the spectacular fossil discoveries of the West. Many of those western finds, as well as the earliest discovered dinosaurs from the East Coast, reside in the great eastern museums, such as the Academy of Natural Sciences of Philadelphia, the Peabody Museum of Natural History at Yale University, the National Museum of Natural History of the Smithsonian Institution, and the American Museum of Natural History.

Dinosaur hunting for these early paleontologists was more than a science. Their zeal was fired by the thrill of illuminating the unknown and changing humanity's very perception of itself, as the great drama of life's evolution over immense stretches of time was slowly revealed. Listen to these eloquent quotations from several of the old pioneers:

> In every quarter, the disemboweled earth yields up its fossil treasures—here of mammal, there of fishes; and in the Cretaceous, east, south and west, a profusion of reptilian remains. . . . Every day is bringing new facts to science, and enlarging the mind of man to fit him for a

more worshipful contemplation of the Creator in the grandeur and exhaustless fertility of design in His works. From the remotest periods a procession of organic forms advances slowly through the misty portals of time. As it proceeds, new things endowed with life and action join the throng, but old forms drop by the wayside, and appear no more forever, but as inanimate dust, marking the "footprints of the Creator."

(Christopher Johnston, 1859)

What a wonderful menagerie! Who would believe that such a register lay buried in the strata? To open the leaves, to unroll the papyrus, has been an intensely interesting though difficult work, having all the excitement and marvelous developments of a romance. And yet the volume is only partly read. Many a new page I fancy will yet be opened.

(Edward B. Hitchcock, 1858)

The magic of early dinosaur paleontology on the East Coast was embodied in many sites and discoveries. There was the bridge in Manchester, Connecticut, built of stone blocks that entombed the missing parts of a dinosaur skeleton for eighty-five years until its demolition finally freed the bones. The Connecticut quarry that produced those stone blocks yielded three dinosaur skeletons during a half-dozen years in the late 1880s, a feat never equaled on the East Coast before or since. In the corridor between Washington, D.C., and Baltimore, so many Early Cretaceous dinosaur fossils were found in the course of surface mining for iron ore in the eighteenth and early nineteenth centuries that the area was nicknamed "Dinosaur Alley."

Today that magic continues, thanks to a group of dedicated collectors who have fashioned a rebirth of dinosaur hunting on the East Coast after more than a half-century of neglect. As author and paleontologist Edwin H. Colbert wrote in 1970, "One of the joys of paleontology is that it will always be an unfinished science, with new treasures ever being yielded up by the earth."

Among the modern dinosaur treasures of the East are two beds of footprints and trackways, 215 million years old, discovered several hundred feet below the surface in an active stone and gravel quarry near Culpeper, Virginia; dinosaur skeletal remains from the steep cliffs eroded by the world's highest tides in Nova Scotia's Bay of Fundy; dinosaur tracks uncovered during construction of a nuclear power plant in Pennsylvania; and bones and teeth recently found in South Carolina, the very first dinosaur evidence from that southern state.

The present vigor of dinosaur hunting on the East Coast is a gift to pale-ontologists and friends of dinosaurs worldwide, offering important clues to dinosaur distribution, evolution, and extinction. We now know that dinosaurs were living in eastern North America soon after they first appeared in South America. Their evolution here paralleled the worldwide unfolding of the dinosaur story and its catastrophic ending 65 Ma, perhaps with the impact of a large asteroid or comet.

The dinosaur bestiary of the East Coast represents fewer species than prime fossil regions elsewhere, but most major dinosaur groups were present. At the dawn of the age of dinosaurs, more than 200 Ma, there were small, primitive plant-eating dinosaurs and track-making meat eaters 6 meters (20 feet) long; by 100 Ma, armored dinosaurs and huge sauropods similar to *Apatosaurus* populated the region. The Mesozoic curtain came down on duck-billed dinosaurs and vicious predators related to *Tyrannosaurus*.

That fossil record has been hard won, fragment by fragment, footprint by footprint, collected by true believers who perhaps appreciate more than their western colleagues the joys of small rewards and the potential value of a single piece of bone. Dinosaur paleontology is a different game on the East Coast, with most fossils recovered from small, ephemeral sites created by the digging, building, and paving of the ever-growing eastern megalopolis. Sadly, many of the most productive historical sites have disappeared beneath overgrowth or concrete and asphalt. Yet the possibility always exists in the East that a great fossil location could appear at any time, after a heavy rain or beneath a backhoe's steel bite.

For collectors of all experience levels and backgrounds, those conditions are a challenge that makes the hunt even more exciting. So much of the East's dinosaur heritage is still waiting to be found, and each new discovery has the potential to add an important piece to the puzzle. Are whole skeletons waiting? Rich bone beds? Fossilized eggs?

Dinosaur hunting is *still* more than a science. Especially when the prey is the dinosaurs of the East Coast.

1

D I N O S A U R I A

Scientists have puzzled over dinosaurs since they were first discovered in the early nineteenth century on the East Coast of North America and in England. When and how did they live? What did they look like? Which regions did they inhabit? Why did they disappear?

Since that time, between three hundred and four hundred dinosaur species have been identified from around the world. Half of that total have been named in just the past twenty years, and some estimates put the species yet to be discovered at three times or more the present count.

Much has been learned about this intriguing group of extinct reptiles. They were extremely successful over the course of the Mesozoic Era and its three periods (from oldest to youngest): the Triassic, Jurassic, and Cretaceous. These animals dominated the land surface of the earth and have been found on every continent, from the North Slope of Alaska to the Antarctic peninsula, from the western Great Plains of North America to the deserts of central and eastern Asia.

There was a constant ebb and flow to dinosaur evolution, making for a rich history that is fascinating to scientists and the public alike. Yet nearly two hundred years after the earliest discoveries, many of the same questions confront today's dinosaur paleontologists. We are still just getting to know these "terrible reptiles."

FIG. 1.1 Sir Richard Owen (1804–1892), founder of Dinosauria in 1842. Owen holds the leg bone of a moa and is wearing robes of professor of comparative anatomy at the Royal College of Surgeons. Oil painting by H. W. Pickersgill 1844. (Natural History Museum, London)

What Is a Dinosaur?

Let's begin at the beginning, with the most fundamental of questions: What is a dinosaur? It was Sir Richard Owen, an English comparative anatomist and paleontologist (fig. 1.1), who coined the name Dinosauria, from the ancient Greek *deinos* ("terrible") and *sauros* ("reptile"). In a paper published in 1842 in the *Proceedings of the British Association for the Advancement of Science*, Owen wrote, "The combination of such characters . . . manifested by creatures far surpassing in size the largest of existing reptiles, will, it is presumed, be deemed sufficient ground for establishing a distinct tribe or suborder of Saurian Reptiles, for which I would propose the name of Dinosauria."

Owen didn't know about some of our most popular dinosaurs, like *Tyrannosaurus, Triceratops,* and *Apatosaurus.* These were discovered with the great explorations of the Western Interior of North America in the late 1800s and early 1900s. But he did have the fossil remains of *Iguanodon, Megalosaurus,* and *Hylaeosaurus,* some of the first dinosaurs to be discovered. And because Owen and other pioneering paleontologists recognized key skeletal features that distinguish dinosaurs from other large extinct animals, we know that *Tyrannosaurus, Triceratops,* and *Apatosaurus* belong to the great family of dinosaurs. Just as humans have genealogical relationships, dinosaurs had groupings of family and relatives, ancestors and descendants, all sharing certain characteristics. This "family" or evolutionary tree helps us to identify major

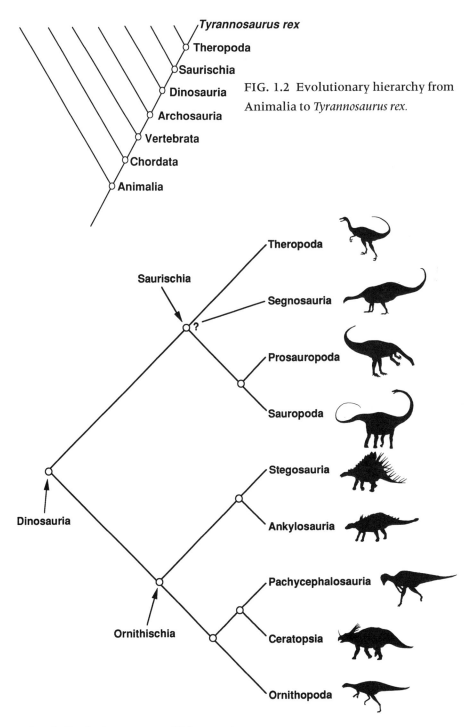

FIG. 1.2 Evolutionary hierarchy from Animalia to *Tyrannosaurus rex.*

FIG. 1.3 Evolutionary tree of Dinosauria.

dinosaur groups and even to understand how dinosaurs are related to modern animals (chapter 10).

The evolutionary trees of dinosaurs and all other living organisms are hierarchical. They consist of large groups containing progressively smaller groups all the way down to genus and species, organized to reflect patterns of common descent. From Animalia—the largest group of organisms whose cells are organized into distinct tissues—to one of the most famous dinosaur species, *Tyrannosaurus rex*, we progress down through Chordata, Vertebrata, and Archosauria, a group that includes living crocodilians and birds (fig. 1.2).

Within Archosauria is Dinosauria, comprising the two major groups of dinosaurs. Ornithischia are the "bird-hipped" dinosaurs, a vast array of plant eaters that includes duck-billed, armored, and horned dinosaurs. The other major dinosaur group, Saurischia, are the "lizard-hipped" dinosaurs. The earliest saurischians gave rise to a number of groups, including Theropoda, and *Tyrannosaurus* evolved within the family of theropod dinosaurs.

The dinosaur names we see most commonly—such as *Tyrannosaurus* and *Velociraptor*—give the genus, often chosen by the person who first described the dinosaur. *Tyrannosaurus* means "tyrant reptile," while *Velociraptor* means "fast stealer." The genus is followed by the species name, which pays homage to a notable researcher, the location where the dinosaur was found, or some characteristic of the dinosaur. So the full names for our two examples are *Tyrannosaurus rex* ("king of the tyrant reptiles") and *Velociraptor mongoliensis* ("fast stealer from Mongolia").

All dinosaurs are either saurischians or ornithischians (fig. 1.3). We have known about each of these groups for as long as we have known about dinosaurs, because the first members of both groups were discovered in England (chapter 4) about 1820: the saurischian *Megalosaurus* ("great reptile") and the ornithischian *Iguanodon* ("iguana tooth").

Saurischians

All saurischian dinosaurs have a lizard-style hip structure: the pubic bone slants down and forward. But so do the hips of many other extinct and living animals, including humans. In fact "lizard hips" are so common among land-living vertebrates (animals with backbones) that we must look to other features uniquely shared by saurischian dinosaurs. Those features include an asymmetrical hand with a big thumb, a long neck, and a back with extra articulations between the vertebrae.

The saurischians comprise four groups: prosauropods, sauropods, theropods, and segnosaurs.

Prosauropods

Prosauropods ("before sauropods") were some of the first saurischians. These plant eaters (fig. 1.4), among the largest dinosaurs of their time, are known from the Late Triassic and Early Jurassic periods in North America, Asia, Europe, Africa, and South America. They measured 7 to 10 meters (23 to 30 feet) in length, and many could reach leaves and fruits as high as 4 meters (13 feet) above the ground when they rose up on their hind legs. In fact prosauropods were the first group of dinosaurs to feed on leaves high in trees.

Best known among prosauropods were *Plateosaurus* ("flat reptile") from Germany, *Massospondylus* ("massive vertebra") from southern Africa, and *Anchisaurus* ("near reptile") from the eastern United States (chapter 6). They all had the prosauropod form: long neck and tail, small and lightly built head, and simple teeth. These animals apparently didn't chew their food but

FIG. 1.4 The prosauropod *Plateosaurus,* from the Late Triassic of Germany and France.

FIG. 1.5 The sauropod *Diplodocus,* from the Late Jurassic of the western United States.

mashed it up in a muscular gizzard full of smooth rocks (known as gastroliths) before it moved on to the stomach for digestion. We know that some prosauropods lived in herds, because their remains are often found in great abundance, almost like a dinosaur graveyard.

Sauropods

Sauropods ("reptile feet") were the closest relatives of prosauropods. Here we find the gigantic *Brachiosaurus* ("arm reptile"), *Diplodocus* ("double beam"), *Apatosaurus* ("deceptive reptile"), and *Seismosaurus* ("earthquake reptile"). These and more than one hundred closely related sauropod species were distributed nearly worldwide, surviving and thriving from the beginning of the Jurassic Period to the end of the Cretaceous.

These dinosaurs fit the popular idea of what a dinosaur should look like: a cross between today's elephant and giraffe, but they were much larger (fig. 1.5), up to 75 tons and 30 meters (100 feet) long. With a small head atop a long neck, and a tail suspended from a rotund body supported on four sturdy legs, these plant eaters probably fed on the leaves of tall trees. Some may even have been able to rear up on their hind legs to feed higher in the forest canopy. Like prosauropods, sauropods had primitive teeth and "chewed" their food in gizzards. They probably lived some of the time in large herds and may have taken care of their young.

Theropods

Theropods ("beast foot") were the supreme hunters of the Mesozoic (fig. 1.6). They were the fiercest of all land-living predators and the largest, at up to 13 meters (43 feet) long. The terrible theropod menagerie included *Tyrannosaurus, Tarbosaurus* ("alarming reptile"), *Allosaurus* ("different reptile"), *Carnotaurus* ("bloody bull"), and *Dryptosaurus* ("wounding reptile").

Although a small number of these giant theropods have enjoyed great attention from scientists and the public, there were others equally menacing. They include *Megalosaurus,* the first discovered of all dinosaurs (1818); *Velociraptor* ("fast stealer"), the terror of *Jurassic Park;* the ostrichlike ornithomimosaurs ("bird-mimicking reptiles"); and the nightmarish oviraptorids ("egg robbers").

The evolutionary history of theropod dinosaurs extends back to the Late Triassic and Early Jurassic periods. The earliest theropods included the small—2 meters (6 feet) long—and frighteningly agile *Coelophysis* ("hollow form") and *Syntarsus* ("fused ankle"). These predatory dinosaurs evolved in a great variety of shapes, sizes, and personalities throughout the remainder of the Jurassic and the Cretaceous. Most had sharp and often serrated teeth. They all walked and ran on their hind legs, a skill that gave them great speed and agility in tracking and making the kill.

Some theropod species, like *Tyrannosaurus* and *Carnotaurus,* had very short forelegs. Others, like *Ornitholestes* ("bird robber") and *Deinonychus* ("terrible claw"), had longer forelegs. This lengthening of theropod forelegs provides an important clue to one of the most intriguing dinosaur questions of all: Are dinosaurs really extinct? (chapter 10).

Theropods appear to have hunted and attacked their prey in a variety of

FIG. 1.6 The theropod *Syntarsus*, from the Early Jurassic of southern Africa.

ways. The larger theropods (*Tyrannosaurus, Carnotaurus,* and *Allosaurus*) probably hunted alone, ambushing or chasing over short distances the likes of duck-billed dinosaurs and young or sickly sauropods, then killing their victims with a powerful bite to the neck. Other theropods like *Deinonychus* and *Troodon* ("wounding tooth") were smaller than their prey and probably hunted in packs as wolves do today when attacking moose.

Segnosaurs

Rounding out saurischian dinosaurs are the segnosaurs (fig. 1.7), a very strange group of plant eaters discovered in 1979 from mid-Cretaceous rocks in central Asia. From the best-known examples—*Segnosaurus* ("slow reptile") and *Erlikosaurus* (named for Erlik, the Lamaist king of the dead)—we know that these dinosaurs resembled a combination of theropods, prosauropods, and ornithischians. They had a small, long head, simple teeth in the back of the jaws but none in the front, a relatively long neck, and broad hips. In addition, the pubic bone was rotated backward, a condition that appears to have evolved twice among dinosaurs: once in segnosaurs and a second time in the great dinosaur group known as ornithischians.

Ornithischians

The ornithischians, a broadly diverse group of plant eaters, are known as the "bird-hipped" dinosaurs because they also have the pubis rotated backward. This characteristic apparently evolved to provide more abdominal room for the fermentation of plant food. Ornithischians share a number of other unique features related to their plant diet: a special "predentary" bone on the tip of the lower jaw; low, triangular tooth crowns on the back teeth; and muscular cheeks to prevent food from falling out of the mouth during chewing.

Ornithischian dinosaurs come in a variety of shapes and sizes, but most can be grouped as stegosaurs, ankylosaurs, pachycephalosaurs, ceratopsians, and ornithopods.

Stegosaurs

Stegosaurs ("roofed reptiles") are some of the most recognizable of dinosaurs (fig. 1.8). They have rows of plates or spines adorning the neck, back, and tail and an additional spine over the shoulder blade. The spines, especially those on the tail and shoulder, are clearly for defense.

But what about the plates along the back? Staggered in two parallel rows, these plates were once thought to have functioned solely in defense. However, recent research suggests that they may have also been used for territorial display and heat regulation. Enlarged plates would have made these an-

FIG. 1.7 The segnosaur *Erlikosaurus,* from the Late Cretaceous of Mongolia.

FIG. 1.8 The stegosaur *Kentrosaurus,* from the Late Jurassic of Tanzania.

imals look much more formidable to competitors. And the plates had porous centers and grooves on the outside, both of which contained a complex network of blood vessels. They would have been well designed as heat radiators and solar panels to regulate body temperature.

These were large animals, though not as large as their plant-eating contemporaries, the gigantic sauropods. Stegosaurs weighed 300 to 1,500 kilograms (660 to 3,300 pounds) and ranged in length from 3 meters (10 feet), like the Chinese *Chungkingosaurus* ("reptile from Chungking") to 7 to 9 meters (23 to 30 feet), represented by the Chinese *Tuojiangosaurus* ("reptile from Tuojiang") and *Stegosaurus* from North America. Stegosaurs had primitive teeth, suggesting that chewing was possible, although not well developed.

Ankylosaurs

Ankylosaurs ("fused reptiles") are the other group of armored ornithischians, and nature lavished on them a full suit of armor. Similar to living armadillos and turtles, ankylosaurs were experts in hunkering down for self-defense (fig. 1.9).

The first-discovered ankylosaur was *Hylaeosaurus* ("Wealden reptile"), one of the original members of Owen's Dinosauria. At the time (1832) it was unclear what these bizarre dinosaurs were all about. But important and spectacular discoveries have since been made in North America and Asia from rocks ranging in age from the Middle Jurassic to the end of the Cretaceous.

We now know that ankylosaurs were covered by a shell-like armor formed of a pavement of bony plates and spines across the neck, back, and tail. These plates also covered the top of the head and often the cheeks and lower jaws.

Some ankylosaurs had a large club at the end of the tail, probably to ward off attacks from predatory dinosaurs. Under the armor the body was broadly rounded, with a large gut for digesting food. The head was broad, short, and equipped with simple, leaf-shaped teeth for chewing plants. Ankylosaurs such as *Sauropelta* ("shield reptile") and *Pinacosaurus* ("board reptile") were about 5 meters (16 feet) long. Others like *Ankylosaurus* ranged up to 9 meters (30 feet). As in stegosaurs, the limbs were short. Slow, sluggish, and low on brainpower—it's a good thing ankylosaurs were so well armored!

Pachycephalosaurs

Pachycephalosaurs ("thick-headed reptiles") were the head butters of the Mesozoic, with a thickened skull and other skeletal modifications that must have served as shock absorbers for head-on collisions (fig. 1.10). This specialized equipment has been found in pachycephalosaurs from their earliest known member—the Early Cretaceous *Yaverlandia* (named for Yaverland Battery, Isle of Wight, England)—through the diverse pachycephalosaurs of the Late Cretaceous of North America and Asia.

We can easily envision male *Stegoceras* ("horned roof"), *Prenocephale* ("sloping head"), or *Pachycephalosaurus* lunging at each other headfirst or bashing at their opponents' sides or thighs. This aggressive behavior may have been

FIG. 1.9 The ankylosaur *Euoplocephalus*, from the Late Cretaceous of western North America.

FIG. 1.10 The pachycephalosaur *Stegoceras,* from the Late Cretaceous of western North America.

accompanied by complex territorial displays, especially in pachycephalosaurs like *Stygimoloch* ("Hell Creek demon"), which had prominent spikes at the back of its head. Most pachycephalosaurs were relatively small, no more than 2 to 3 meters (7 to 10 feet) long. The largest, *Pachycephalosaurus* itself, probably measured up to 8 meters (26 feet) long. All were peaceful plant eaters, except when crashing headlong into each other.

Ceratopsians

The horned dinosaurs known as ceratopsians ("horned faces") roamed many of the same habitats as pachycephalosaurs. These fascinating dinosaurs (fig. 1.11) left one of the most outstanding fossil records of any group. They were particularly diverse and well preserved in the latest Cretaceous of North America. Yet the first members of Ceratopsia are known almost exclusively from central Asia earlier in the Cretaceous. These early ceratopsians were 2 to 3 meters (7 to 10 feet) long and often hornless. They included *Psittacosaurus* ("parrot reptile"), *Protoceratops* ("first horned face"), and *Bagaceratops* ("small horned face").

Ceratopsian dinosaurs reached their evolutionary peak toward the end of the Cretaceous. Ranging from 6 to 9 meters (20 to 30 feet) long and weighing 6 to 7 tons, they had very large skulls adorned with an impressive variety of horns and a broad shield (called a "frill" by dinosaur paleontologists) extending back over the neck and shoulders. *Triceratops* ("three-horned face") and *Chasmosaurus* ("opening reptile") had a long horn over each eye and a smaller one over the nose. *Centrosaurus* ("spur reptile") and *Styracosaurus* ("spike reptile") had smaller eye horns and a long nose horn. The most unusual ceratopsian was *Pachyrhinosaurus* ("thick-horned reptile"), which sported no horns but had a large block of roughened bone over the nose and eyes.

Ceratopsians started out walking on their hind legs but soon evolved a four-footed stance. Both forelimbs and hind limbs were powerfully built, and ceratopsians were fast runners, perhaps up to 14 meters a second (31 miles an hour). Imagine a *Triceratops* with its pointy end facing forward and bolting at a threatening *Tyrannosaurus!* Yet the horns and frills may also have functioned in mating display, ritualized combat between males, and other social behavior. These are the same functions horns serve in modern antelope, cattle, giraffe, rhino, and gazelle.

Horned dinosaurs were among the most successful of plant eaters in the Mesozoic, and we can learn why by looking at the jaws. All ceratopsians snapped off leaves from trees and shrubs with their powerful parrotlike beaks and chewed with a slicing and dicing action of their strong teeth and jaw muscles. Fossil evidence for at least one ceratopsian, *Psittacosaurus*, reveals that the food was processed a second time in a gizzard, much as in prosauropods and sauropods.

FIG. 1.11 The ceratopsian *Styracosaurus,* from the Late Cretaceous of western North America.

FIG. 1.12 The ornithopod *Iguanodon*, from the Early Cretaceous of Europe and the western United States.

Ornithopods

Ornithopods ("bird feet"), along with ceratopsians, were among the most prolific of the plant eaters, but ornithopods had a much longer history and a much wider range of shapes and sizes than the horned dinosaurs. Many ornithopods are known from well-preserved fossils, including those of teenagers, juveniles, hatchlings, and even embryos. From these specimens, and from abundant egg nests, we are learning more about parental care of offspring, group nesting, and the rapid growth rates that appear to characterize ornithopods.

From the Early Jurassic come the small—2 meters (7 feet) long—tusked ornithopods known as heterodontosaurids, such as *Lycorhinus* ("wolf snout") and *Heterodontosaurus* ("different-toothed reptile") from southern Africa. From the Late Jurassic to the end of the Cretaceous, the ornithopods called hypsilophodontids and iguanodontians predominated nearly worldwide. Hypsilophodontids were fast running and gazelle-like, no more than 2 meters (7 feet) long. They included *Hypsilophodon* ("high-crested teeth") and *Orodromeus* ("mountain runner").

Iguanodontians, larger and more horselike, are probably the most familiar of all ornithopods. Especially well known are the hadrosaurids, or duck-billed dinosaurs. *Iguanodon* ("iguana tooth") is one of the first-discovered dinosaurs (1822) and another charter member of Owen's Dinosauria (fig. 1.12). This 10 meter (33 feet) long ornithopod from the Early Cretaceous of Europe had a long, straight tail, a stocky body, and huge spikes on its thumbs.

Most hadrosaurids, such as *Maiasaura* ("good mother reptile") and *Gryposaurus* ("hooked-nose reptile"), look like a cross between a duck and an ox. Their beaks were probably used to harvest great quantities of leaves and other vegetation. Some hadrosaurids had peculiar adornments atop their heads, such as the hollow crests of *Corythosaurus* ("helmet reptile") and *Parasaurolophus* ("like *Saurolophus*"). Scientists theorize that these bizarre headdresses were useful both for their visual impact and to enhance vocal range and repertoire, especially for the hollow-crested duck-billed dinosaurs.

Ornithopods were the best chewers of the Mesozoic. Like the great majority of ornithischians, they had muscular cheeks and strong and abundant upper and lower teeth that formed flat grinding surfaces to rip apart favorite leaves, stems, and fruits. Like ceratopsians, ornithopods appear to have been the earliest plant eaters to take advantage of the new food source provided by the first flowering plants (angiosperms) in the Early Cretaceous.

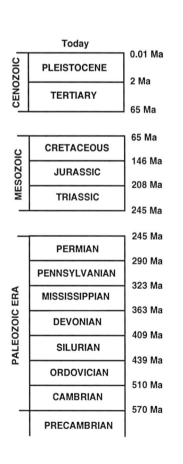

FIG. 2.1 Geologic time scale.

2

AN EVOLVING WORLD

Before the story of dinosaurs on the East Coast of North America can be told, we must hear the long, slow tale of Earth: the dance of the continents, the ebb and flow of oceans, the rise of mountains and their steady erosion, the origin of rocks and how they preserve the record of ancient life.

And to put into perspective the billions of years that have elapsed since Earth was born, we need the relative scale of great spans of time (fig. 2.1) that includes the Mesozoic Era—the age of dinosaurs—and its three periods, the Triassic, Jurassic, and Cretaceous.

The Jigsaw Puzzle

Based on the pioneering voyages of such great explorers as Christopher Columbus, Ferdinand Magellan, Vasco da Gama, Amerigo Vespucci, and James Cook, even some of the earliest world maps showed the continents shaped like pieces of a giant puzzle made to fit closely together. The contours of the east coasts of North and South America clearly corresponded to the western coastlines of Europe and Africa. But how could that be? In the prevailing view, Earth was a stable place. Continents were "solid as a rock."

A controversial explanation by German meteorologist Alfred L. Wegener appeared in 1915. In *The Origin of Continents and Oceans,* Wegener suggested that the continents were originally joined in a single huge landmass that he called Pangaea, Greek for "all land." This supercontinent gradually pulled apart, and the individual continents moved to their present positions.

Wegener's theory of "continental drift" was greeted with great skepticism, even hostility, partly because there was no known mechanism to explain continental movement. In addition, Wegener made a number of unfortunate assumptions. He claimed that drift was occurring at the galloping rate of 36 meters (118 feet) a year and that it was caused by the centrifugal force of the spinning earth and the gravitational tug of the sun and moon.

The accepted explanation at that time for the formation of continents was the "wrinkled apple" theory of geologist James Dwight Dana of Yale University. Dana believed that all of Earth's geological features, including mountains and oceans, were the result of cooling and contraction of the planet from an original molten state.

It was nearly forty years before scientific evidence supported Wegener's concept of drifting continents. Geologist Harry H. Hess of Princeton University pieced together the modern theory in his 1962 publication "History of the Ocean Basins." The continents ride on huge "plates" of the earth's crust, Hess wrote. These plates are slowly rearranging the landscape of Earth's surface—at about 2.5 centimeters (1 inch) a year—and they are continually recycled. New crust forms as molten rock rises from a 64,000 kilometer (40,000 mile) system of mountainous midocean ridges. Old crust is bent down, or "subducted," into the molten mantle through deep oceanic trenches at continental margins.

Further proof for this process of "plate tectonics" has been accumulating for decades. Seismic imaging—the use of reflected sound waves to form a picture of a geologic feature—and radar mapping of the ocean floor from satellites have clearly revealed the deep trenches and midocean ridges. Another major clue involves paleomagnetism. When certain rocks were still molten, their iron particles aligned to Earth's magnetic pole. After the rock had solidified, that ancient magnetic bearing was literally set in stone. Researchers can now "read" the rocks to determine how much movement has occurred relative to the magnetic pole over millions of years.

As Wegener correctly suggested, the world's landmasses were once joined together in a supercontinent. Pangaea extended some 18,000 kilometers (11,000 miles) along the equator into both the Northern and Southern Hemispheres, covering nearly two-fifths of Earth's surface. Later, driven by the same forces that created the supercontinent, the Pangaean landmasses of

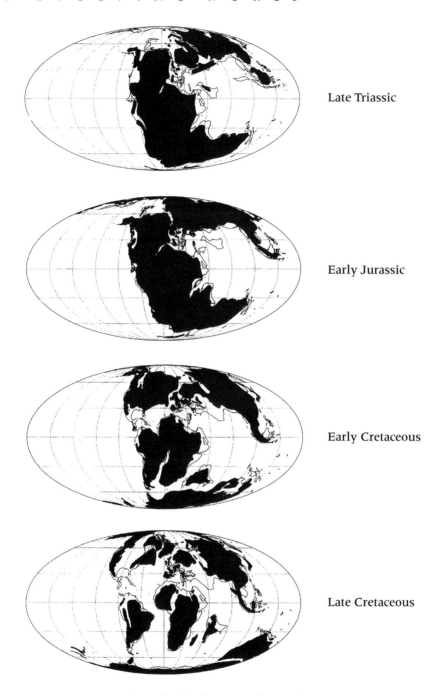

Late Triassic

Early Jurassic

Early Cretaceous

Late Cretaceous

FIG. 2.2 Position of Earth's landmasses during the Mesozoic Era. (After A. G. Smith, D. G. Smith, and B. M. Funnell, *Atlas of Mesozoic and Cenozoic Coastlines* [Cambridge University Press, 1994])

Laurasia (North America, Europe, and Asia) and Gondwana (South America, Africa, India, Antarctica, and Australia) pulled apart to form the present continental configuration (fig. 2.2).

The suspected engine of Wegener's continental drift is heat from the earth's core. A cross section of our planet reveals a solid rock crust for the upper 30 kilometers (18 miles), atop a 3,000 kilometer (1,800 mile) semimolten mantle and 4,800 kilometers (3,000 miles) of core, molten on the outside, solid at the center. Geologists believe that the conveyor-belt system of moving plates is propelled by convection currents circulating in "cells" within the mantle.

Although Wegener postulated only one supercontinent in Earth's history, evidence suggests that Pangaea may have been preceded by two or three additional episodes of continental convergence. At an average drift rate of 6 centimeters (2.5 inches) a year, for example, the continents could have traveled around the earth four times in the past two billion years!

The breakup of Pangaea is still under way, and seven major and eighteen minor plates have been identified, ranging in thickness from 50 to 145 kilometers (30 to 90 miles). The major plates are the African Plate, Eurasian Plate, Indo-Australian Plate, Antarctic Plate, South American Plate, North American Plate, and Pacific Plate.

The huge African Plate is steadily pulling away from the retreating North American and South American Plates. This continental split created the East Coast of North America and opened the Atlantic Ocean. It formed the landscape where the dinosaurs of the East Coast roamed. And it profoundly influenced the depositional environments that preserved their footprints and bones.

How Old?

The concept that Earth is immensely old and has dramatically evolved over billions of years is one of the major accomplishments of science. In the Western world, where the science of geology had its beginnings, religious doctrine originally taught that our planet was young and unchanging. In 1650 James Ussher, Archbishop of Armagh and Primate of All Ireland, concluded in his *Annals of the Ancient and New Testaments* that heaven and earth had been created less than six thousand years before, on Sunday, October 23, 4004 B.C. Dry land emerged that Tuesday, and humans took center stage on Friday, October 28.

Following the publication in 1859 of Charles Darwin's *On the Origin of Species*, debate over his controversial theory of evolution gave new urgency to questions of Earth's age and the origins of life. Clearly, the evolutionary processes Darwin proposed couldn't have taken place in Ussher's six thousand years, or even in the considerably longer periods suggested by some later theorists. Age estimates were steadily revised, always pointing to an earth that was much older than previously thought.

The Scottish physicist Lord Kelvin—who remained "on the side of the angels" in the evolution debate—applied the most elegant logic of nineteenth century science to the question of Earth's age. Unaware of the sun's self-sustaining nuclear power source, he thought that its heat output was rapidly declining and must have been too intense within recent geologic time for life to exist on Earth. Based on those assumptions and others, Kelvin concluded in 1897 that Earth had been habitable for a mere 20 million to 40 million years.

Although some scientists in the late 1800s considered Kelvin's estimate much too low, his reputation and their lack of evidence kept dissent to a minimum. They had to await the debut of a remarkable new tool that revolutionized the science of age dating and expanded geologic time to nearly unimaginable limits.

Radioactive Dating

The death blow to Kelvin's careful calculations came from a newly discovered physical property known as radioactivity. First detected in 1896 by the French physicist Antoine-Henri Becquerel, it was soon identified in several other elements by the husband-and-wife team of Pierre and Marie Curie. In 1906 English physicist H. J. Strutt concluded that heat was being generated at Earth's core by the radioactive decay of elements.

Radioactivity first revealed its age-dating potential when it became clear that radioactive "isotopes"—unstable forms of certain elements—decay at a known rate to stable by-products. Each has a unique "half-life": the time for half of the "parent" isotope to convert to a stable "daughter" product. This halving takes place at the constant rate again and again until the unstable parent has completely decayed. By measuring how much daughter exists relative to the parent isotope, scientists can precisely determine the passage of time.

The initial discoveries leading to a system of "radiometric" dating were made by chemist Bertram B. Boltwood of Yale University, who found that radioactive isotopes of uranium decay to lead. In 1907 Boltwood proved that the lead (daughter) to uranium (parent) ratio in minerals increased with the age of the mineral, and he made the first modern estimates of Earth's age. The results, although crude by today's standards, were astounding. Boltwood dated some of the oldest known rocks at more than 1.6 billion years.

Radiometric analysis is a form of *absolute* dating, meaning that the age of a particular rock can be determined independent of its surroundings. Approximately twenty long-lived isotopes have been identified for radiometric dating purposes, though only a handful are abundant enough in nature for practical use. Uranium-lead is one of the mainstays: uranium 238 has a half-life of 4.5 billion years; the half-life of uranium 235 is 713 million years. Others include potassium-argon (1.3 billion years) and thorium-lead (14 billion years).

Most rocks that can be dated by radiometric methods are nonsedimentary, such as granite and volcanic lava. But fossils are found only in sedimentary rocks, so paleontologists must understand the evolution of rocks and their geologic relationships to determine their relative age. *Relative* dating determines the age of a naturally buried object by noting its position relative to rocks or fossils of known age.

The Nature of Rocks

Although the world exhibits a wide range of types, shapes, and colors of rock, each belongs in one of three categories based on origin:

Igneous rocks, such as granite and basalt, crystallize from molten magma rising from the hot mantle. This rock type may have formed as an intrusion into other rocks deep underground, or it may be extrusive—deposited on Earth's surface as a lava flow.

Metamorphic rocks have been altered (but not melted) by high heat or pressure into crystalline rock forms. In metamorphism shale becomes slate, sandstone becomes quartzite, and limestone becomes marble.

Sedimentary rocks—including sandstone, limestone, and coal—are deposited as layers of mud, clay, gravel, or organic remains of plants and animals. These layers are converted to rock by the pressure of overlying strata. Sedimentary rocks constitute about 75 percent of all exposed rocks on Earth.

In the nomenclature of geology, the smallest mappable rock unit is called a *formation*. Names of formations consist of two parts: a geographic name taken from the locale where a formation was first identified, followed by the word "Formation" or the rock type, such as "Shale" or "Sandstone." For example, the first nearly complete dinosaur skeleton in the world was described in 1858 from a rock unit in Haddonfield, New Jersey. Known as the Woodbury Formation, or Woodbury Clay, the unit was named for the town of Woodbury, New Jersey.

Smaller rock units are called *members,* and two or more formations may be classified together as *groups.* In some cases groups are part of larger units known as *supergroups.* The Newark Supergroup on the East Coast of North America is a series of Triassic/Jurassic formations exposed from South Carolina to Nova Scotia and containing numerous dinosaur bones and footprints (chapter 3). *Strata* are layers of sedimentary rock that were originally deposited as sediments on floodplains and at the bottom of lakes and oceans. These layers may have been dramatically tilted and folded in *orogenies:* episodes of mountain building caused by tremendous forces associated in many cases with the collision of continents.

Stratigraphy is the study of strata and their origin, composition, distribution, and succession through time. Geologists correlate strata from different regions to determine geologic relationships and to understand processes that were happening at the same time in different areas. Some of the earliest clues to continental drift involved comparing similar rock layers on opposite sides of the Atlantic Ocean, in regions that were once unified as part of Pangaea.

Geologists and paleontologists use every available clue to understand the vanished world of dinosaurs. Rocks reveal processes of geologic change and past environment, and they preserve fossils, the only record of ancient life and its evolution. Rocks help establish the age and identity of fossils through their relative position in layers of known age and composition. In turn, fossils illuminate the nature and age of rocks.

Fossil Treasure

What is a fossil? The word derives from the Latin *fossilis,* meaning "something dug out of the ground." Once broadly interpreted as anything of any age found buried in the earth, the term is now restricted to the remains or traces of organisms from approximately 10,000 years or more into the geologic past.

Fossils take many forms. They include preserved pieces of animals or plants, such as bones, teeth, shells, wood, and stems; footprints, known as *ichnites;* impressions of skin or leaves; eggs, or *ooliths,* very rarely with embryos inside; nests and burrows; plant pollen and spores; and even excrement, called *coprolites.*

To become fossils, dead plants or animals must be rapidly buried. Left exposed on the earth's surface, organic remains are eaten by other animals or quickly decomposed by bacteria and the elements. Even bone turns to dust. As a result, the most abundant fossils are those of marine organisms, preserved on the bottom of ancient seas. Less common are the fossils of plants and animals that lived in or near lakes, streams, and rivers. The rarest fossils are of terrestrial animals, such as dinosaurs, that were preserved when they died near a body of water or their bones were washed into the sea. Extremely dry conditions, such as volcanic ash or sand, may also preserve animal fossils.

Most dinosaur fossils, including entire skeletons, have been preserved through processes that hardened the buried bones with minerals over millions of years. In one, known as *permineralization,* the internal organic structure of bone, shell, or wood remains intact, but holes and tiny pores are infiltrated with minerals such as calcite, iron sulfide, or silica from groundwater. This gives hollow bones strength not to be crushed as layers of sediment accumulate above them. Another fossilization process, *petrification,* turns all of the organic material to stone, which may destroy its cellular structure. In contrast, teeth—the hardest of all skeletal parts—often survive the eons unchanged.

In some cases, after the organic remains are embedded in sediment they are dissolved by groundwater, leaving a hollow known as a *mold* in the exact shape of the bone. Molds of thin objects, such as plant leaves or reptile skin, are called *imprints.* A mold that was later filled with minerals is a *cast* of the original animal or plant remains.

Another kind of fossil preserves evidence of animals or plants other than organic remains. These *trace fossils* include dinosaur footprints left on muddy lakeshores more than 200 Ma. Many dinosaur tracks and other reptile tracks have been found on the East Coast of North America, particularly in the Connecticut Valley. Tracks can reveal valuable clues to the size and behavior of animals: whether they walked on two feet or four, how fast they moved, if they lived in groups, and what environments they preferred.

Among the rarest of fossils are those of animals preserved whole, with soft parts still largely intact. Paleontologists identify three special environments that can produce this form of fossilization: amber, ice, and oil. Amber is the

fossilized resin or sap of ancient plants, and insects attracted to the resin became trapped and were entombed when it turned to solid amber. The permanently frozen ground of the Arctic has yielded the preserved remains of
mammoths that lived 30,000 years ago, and whole carcasses of woolly rhinoceros have been collected from oil seeps in Poland. The famous La Brea tar
pits near Los Angeles entombed such Pleistocene animals as mammoths,
horses, bison, wolves, and saber-toothed tigers.

Fossils have played a major role in the understanding of Earth's geology,
as well as the evolution of life. In 1799 English engineer William Smith first
recognized what later came to be called the "law of faunal succession." This
principle states that strata may be identified by the fossils they contain. Certain *index fossils*—often pollen or spores—from strata of known age can be
used to identify rock layers of similar age on opposite sides of a valley, or on
opposite sides of the world.

The Big Picture

> The great earth itself is rent with commotions. A continent sinks below
> or rises above the mighty deep. But onward, still onward, grandly moves
> the pageant, evermore in the growing light. The ground trembles be-
> neath its heavy tread—its pulse beats the "seconds of eternity," and its
> voices shake the air.
>
> Christopher Johnston, 1859

Earth is now known to be an ancient planet indeed, even more ancient
than Boltwood's 1.6 billion years. The best current estimate for its age is 4.6
billion years. Our planet condensed with the sun and the rest of the solar system from a swirling nebula of gas and dust, a process perhaps set in motion
by shock waves from a nearby exploding star.

The 4.6 billion year estimate is largely based on radioactive dating of me-
teorites, bodies that formed at the same time as Earth and still preserve traces
of the primordial material. The oldest exposed rocks found so far are in western Greenland and Africa, approximately 3.7 billion to 3.9 billion years old.
That leaves a gap in the early record of at least 700 million years. Scientists
speculate that intense meteorite bombardment of the molten Earth continued until about 3.9 billion years ago (Ba), when the crust may finally have
cooled enough for rocks to form.

The history of geologic time on Earth has been refined by geologists over
the past two hundred years into a system of "time-stratigraphic" units called

(from largest to smallest): eons, eras, periods, epochs, and ages. Epochs are the "Early," "Middle," and "Late" designations that appear before period names, as in "Early Triassic." Geologic convention dictates the use of "Lower," "Middle," and "Upper" when referring to rock units, as in "Lower Triassic."

By far the largest unit of geologic time is also the oldest. Known as the *Precambrian* ("before the Cambrian"), this interval encompasses more than 85 percent of Earth's evolutionary history, from 4.6 Ba to 570 Ma. The oldest forms of life found on Earth are marine fossils of tiny bacteria and blue-green algae from South African rocks approximately 3.2 billion years old. The first evidence of multicellular organisms dates from about 750 Ma: fossilized burrows and trails made in the seafloor by wormlike invertebrates.

From 570 Ma to the present, geologic time is encompassed by the *Phanerozoic Eon*, derived from a Greek phrase that means "visible life." The Phanerozoic is divided into three eras (from oldest to youngest): the Paleozoic ("ancient life"), the Mesozoic ("middle life"), and the Cenozoic ("recent life").

The oldest Phanerozoic era, the *Paleozoic*, spans 325 million years and is divided into seven distinct periods of unequal length (from oldest to youngest): the Cambrian (570 Ma to 505 Ma), Ordovician (505 Ma to 438 Ma), Silurian (438 Ma to 408 Ma), Devonian (408 Ma to 360 Ma), Mississippian (360 Ma to 320 Ma) (fig. 2.5), Pennsylvanian (320 Ma to 286 Ma), and Permian (286 Ma to 245 Ma).

The Paleozoic was the vast stage for the great flowering of life on Earth. Marine invertebrates flourished and diversified in the Cambrian. The first vertebrates—primitive jawless fishes—arrived in the Ordovician. The first land animals appeared in the Silurian, in the form of scorpionlike arachnids, and the first tetrapods ("four legs") in the Late Devonian. The first reptiles evolved at the close of the Pennsylvanian.

Jumping ahead to the youngest Phanerozoic era, the *Cenozoic* is divided into the Tertiary Period (65 Ma to 2 Ma) and the Quaternary Period (2 Ma to the present). The Tertiary is further subdivided into five epochs (from oldest to youngest): the Paleocene (65 Ma to 57 Ma), Eocene (57 Ma to 34 Ma), Oligocene (34 Ma to 23 Ma), Miocene (23 Ma to 5 Ma), and Pliocene (5 Ma to 2 Ma). The Quaternary comprises two epochs, the Pleistocene (2 Ma to 10,000 years ago) and the Holocene (10,000 years ago to the present).

The Cenozoic is the age of mammals. This diverse and highly adaptable group bloomed after the mass extinction of dinosaurs at the end of the Mesozoic (chapter 10). The last part of the Pleistocene, some 15,000 years ago, is

FIG. 2.3 The spectacular Sideling Hill road cut in western Maryland exposes marine sediments 360 to 340 million years old from the Mississippian Period of the Paleozoic Era. (John Thrasher, Above All Photo)

marked by the arrival on the East Coast of the most successful of mammals: *Homo sapiens.* And approximately 13,200 years later, this inquisitive animal began discovering the remains of great extinct reptiles that had roamed eastern North America millions and millions of years before, in the Mesozoic Era.

Age of Middle Life

The Mesozoic lasted 180 million years, approximately 4 percent of Earth's long history, and its three periods are the familiar framework for the dinosaur story. The Triassic Period (245 Ma to 208 Ma) was named in 1834 for the threefold division of Triassic rock sediments in southern Germany. The Jurassic (208 Ma to 145 Ma) was named for limestones in the Jura Mountains of France. And the Cretaceous (145 Ma to 65 Ma) was named for white chalks (*creta* in Latin) exposed in Belgium, northern France, and England.

The Mesozoic world was far different from that of today. The East Coast of

North America was stranded deep within Pangaea when the era began. The moon appeared bigger in the sky, since it was closer to Earth than now. Because our planet was also spinning faster, the days were shorter: 22 hours and 45 minutes.

But conditions must have been just right. Over the course of the Mesozoic, dinosaurs would emerge as some of the most successful land vertebrates in Earth's history. And their story begins on the East Coast.

3

THE MESOZOIC EAST COAST

The dinosaur story on the East Coast of North America is told in the rocks of the Mesozoic Era. Once the muds and clays dinosaurs walked and died on, those rocks now preserve our only clues that these extinct reptiles existed.

Triassic Beginnings

During most of the Triassic Period, the future East Coast was landlocked deep within the interior of the supercontinent Pangaea, crushed against Europe and the west coast of Africa at a latitude much closer to the equator than today. The climate was warm and intensely seasonal, owing to the influence of the huge unbroken landmass on atmospheric and ocean circulation patterns. Ponds and regional lakes advanced and retreated as rainy and dry cycles alternated. Plant life was dominated by tall conifers, ginkgoes, ferns, horsetail rushes, and cycads. Flowering plants, known as angiosperms, wouldn't appear until the Early Cretaceous.

The early Mesozoic animal world was in transition. A great extinction at the end of the Paleozoic Era had killed off most species of marine organisms, and some land animals and plants were also severely affected. Beginning in Triassic times, there was a gradual rebirth of the ocean invertebrates, especially ammonites, distant relatives of today's pearly nautilus. On land, primitive mammal-like reptiles (therapsids) were in decline, and the first true mammals appeared. More important for our story, it was at this time that the archosaurian reptiles began their advance (chapter 10).

Near the end of the period, in the Late Triassic, the first evidence of dinosaurs appears in rocks approximately 230 million years old. On the East Coast, the earliest footprints and fossil bones of dinosaurs are preserved in a system of ancient rift basins now exposed in a narrow, 1,600 kilometer (1,000 mile) long band from Nova Scotia to the Carolinas.

Newark Supergroup

Pangaea began to break apart near the end of the Triassic, as North America separated from western Africa and Europe. Earth's crust thinned and stretched along a rift axis parallel to the present-day East Coast, forming a series of deepening troughs, or fault-bounded basins. Many of these basins developed deep freshwater lakes, but only one was "successful" in rifting widely and deeply enough to be flooded by oceanic waters. This was the beginning of the Atlantic Ocean, which continues to grow today as the North American and African plates are pushed steadily away from each other (chapter 2).

Over millions of years, the other, smaller rift basins were filled with thousands of feet of gravel, sand, silt, and clay eroded from highlands to their east. These sediments are "interbedded" with igneous rock and basalt lava flows from volcanic activity that accompanied the breakup of Pangaea and the birth of the Atlantic. For the most part the sediments were deposited on broad plains by flooding rivers. Now exposed as reddish sandstones, these "redbeds" preserve many dinosaur footprints. Localized pockets of black or gray shale were laid down in shallow ponds, lakes, and swamps during this period. The bones and teeth of dinosaurs are most likely to be found in these rock deposits.

Named the Newark Supergroup in 1978 by Paul E. Olsen (chapter 9), the rocks of the rift basins are exposed in an irregular line along the East Coast. As recently as the 1970s, all the rocks were thought to be Upper Triassic in age. But careful correlation of fossil spores, pollen, and fishes by Olsen and others has established an Early Jurassic age for many of the rock formations in the Newark system.

Current studies have identified up to thirty exposed Newark Supergroup basins in nine states and the Canadian maritime province of Nova Scotia

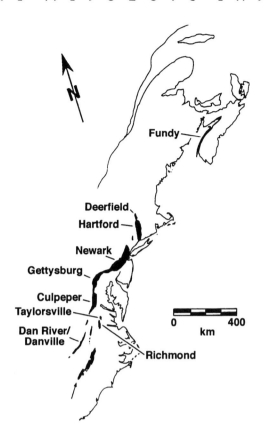

FIG. 3.1 Map of major Newark Supergroup basins, East Coast of North America (after Olsen 1978).

TABLE 3.1   Newark Supergroup Basins (from south to north)

Crowburg Basin—South Carolina
Wadesboro Basin—South Carolina, North Carolina
Ellerbe, Sanford, Durham, and Davie County Basins—North Carolina
Dan River/Danville Basin—North Carolina, Virginia
Scottsburg, Randolph, Roanoke Creek, Briery Creek, Farmville, Richmond, Flat Branch,
   Deep Run, and Taylorsville Basins—Virginia
Scottsville and Barboursville Basins—Virginia
Culpeper Basin—Virginia, Maryland
Gettysburg Basin—Maryland, Pennsylvania
Narrow Neck of the Gettysburg/Newark Basin—Pennsylvania
Newark Basin—Pennsylvania, New Jersey, New York
Pomperaug Basin and Cherry Brook Outlier—Connecticut
Hartford Basin—Connecticut, Massachusetts
Deerfield, Northfield, and Middleton Basins—Massachusetts
Fundy Basin—New Brunswick, Nova Scotia
Chedabucto Basin—Nova Scotia

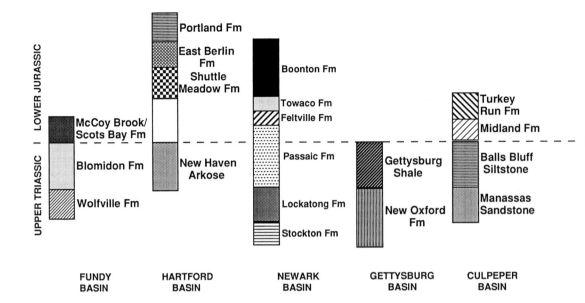

(fig.3.1, table 3.1). Buried extensions of these rift basins continue south into Florida and north as far as Newfoundland.

Deposition of the oldest Newark Supergroup sediments apparently began concurrently in the southern basins and in Nova Scotia as early as 235 to 230 Ma. In contrast, the youngest Newark Supergroup sediments were laid down in the northern basins approximately 180 Ma during the Early Jurassic, as North America and Africa were finally parting company and deposition in these basins probably ceased (fig. 3.2).

During the Early Jurassic Period, climate was greatly influenced by the on-going Pangaean split. The future East Coast, no longer blocked from mois-ture-bearing ocean winds, grew warmer and more humid. The steadily widening Atlantic Ocean also moderated climate shifts. Another great ex-tinction, at the boundary of the Triassic and Jurassic, killed off about half of the groups of terrestrial reptiles. But the dinosaurs survived to continue their rise to dominance of all land vertebrates.

Connecticut Valley

The very first dinosaur fossils from the East Coast rift basins were found in the Connecticut Valley. This ancient trough follows the present-day water-shed of the Connecticut River. It extends 178 kilometers (105 miles) from

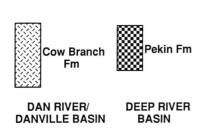

FIG. 3.2 Correlation of Upper Triassic–Lower Jurassic geologic formations in the Newark Supergroup, East Coast of North America.

Northfield, Massachusetts, near the New Hampshire line, to Long Island Sound at New Haven, Connecticut.

The old Connecticut Valley was the creation of a geologic feature known as the Eastern Border Fault. Powerful earthquakes along the fault dropped the valley's floor. Rivers carrying eroded sediments from the mountains to the east eventually deposited up to 4,900 meters (16,000 feet) of sand, gravel, silt, and clay, now turned to rock. Lakes, some as large as 96 kilometers (60 miles) long and 16 kilometers (10 miles) wide, covered the valley floor. Although coal beds formed in subtropical swamps in rift basins to the south, no coal deposits are known from the Connecticut Valley.

The Connecticut Valley rocks of most importance to dinosaur hunters are the sandstones. Oxidized to reddish brown over millions of years, this distinctive stone became the foundation of a bustling quarrying industry in the 1800s. Connecticut Valley "brownstone" was in great demand by architects of elegant Victorian-era buildings in New York, Boston, and other New England cities. The industry peaked in the late nineteenth century, and the closing of these quarries cut off a major source of dinosaur fossils on the East Coast.

The Connecticut Valley is made up of two major Newark Supergroup basins, the Hartford Basin to the south and the smaller Deerfield Basin in the north. They are separated in central Massachusetts by strong Paleozoic bedrock that has resisted erosion during the 200 million years since the basins were created.

The Hartford Basin is approximately 120 kilometers (75 miles) long. Named for Hartford, Connecticut, it extends from Long Island Sound, New York, to the Hadley–South Amherst area of Massachusetts. It is divided into seven formations, from oldest to youngest: the New Haven Arkose (the only Triassic formation), Talcott Basalt, Shuttle Meadow Formation, Holyoke Basalt, East Berlin Formation, Hampden Basalt, and Portland Formation.

The Deerfield Basin—named for Deerfield, Massachusetts—is about 30 kilometers (19 miles) long and located entirely in Massachusetts, from Hadley–South Amherst in the south to near Northfield in the north. The Deerfield formations are, from oldest to youngest: the Sugarloaf Arkose (the only Triassic formation), Deerfield Basalt, Turners Falls Sandstone, and Mount Toby Formation.

The basalt formations in these basins, which preserve no fossils, represent major lava flows during the Early Jurassic and are visible today in such features as the 190 million year old Holyoke Range in Massachusetts. The lava flows altered valley drainage patterns to promote flooding and lake expansion, with a subsequent increase in sedimentation that boosted chances for fossil preservation.

No dinosaur fossils have been found in the New Haven Arkose or the Sugarloaf Arkose, the oldest of the Connecticut Valley Newark sedimentary units. The New Haven Arkose sediments began accumulating in the Late Triassic some 200 Ma. This formation has produced fossils of a variety of reptiles, including the crocodile-like phytosaur *Clepsysaurus* from Simsbury, Connecticut. Also from the New Haven Arkose are a skull and skeleton of *Hypsognathus*—a small, plant-eating predecessor to living reptiles—found in Meriden, Connecticut, and a skin impression of the armored aetosaur *Stegomus* from the Freeman Clark Quarry in Fair Haven, Connecticut, described in 1896 by Othniel C. Marsh (chapter 4).

The remaining Hartford Basin formations and their Deerfield Basin counterparts—all of Early Jurassic age—preserve the entire record of dinosaurs of the Connecticut Valley (chapter 6).

Newark Basin

The Newark Basin is the largest of the East Coast rift basins. It stretches 225 kilometers (140 miles) across three states, from Rockland County, New York, through New Jersey to northeast Lancaster County, Pennsylvania. Named for Newark, New Jersey, the basin attains a maximum width of 51 kilometers (32 miles) along the Delaware River and covers an estimated 7,000 square kilometers (3,000 square miles).

Evolution of the Newark Basin was similar to that of the Hartford Basin in the Connecticut Valley. Sediments eroded from the eastern heights accumulated to a thickness of up to 8 kilometers (5 miles), interbedded with lava flows and igneous rock (fig. 3.3). As in the Connecticut Valley, red sandstones are the landscape of reptile footprints. The fossil remains of dinosaurs, fishes, and plants are preserved in rocks that were once black and gray muds laid down on the shores and the bottoms of lakes that cycled in size with climatic changes. The largest of these lakes, known as Lake Newark, existed for millions of years over much of the basin in New Jersey.

The thickest continuous accumulation of sediments in the entire Newark Supergroup is in New Jersey. From oldest to youngest, the sediments include Upper Triassic units—the Stockton Formation, Lockatong Formation, and most of the Passaic Formation—and Lower Jurassic units: the uppermost Pas-

FIG. 3.3 The Palisades in New York State, a 200 million year old lava formation in the Upper Triassic–Lower Jurassic Newark Basin. (Published with the permission of the New York Geological Survey/New York State Museum)

saic Formation, Orange Mountain Basalt, Feltville Formation, Preakness Basalt, Towaco Formation, Hook Mountain Basalt, and Boonton Formation. The two oldest layers of the Newark Basin—the Stockton and Lockatong—have no counterparts in the Hartford Basin, but the remaining strata correspond closely in age with the Hartford rock layers.

The basalt formations help date the Newark sedimentary units. The Orange Mountain Basalt is the major dividing line between the Triassic and Jurassic periods in the Newark Basin, although the uppermost layer of the Passaic Formation has yielded a few Early Jurassic fossils. In north-central New Jersey the basalts are prominently exposed in the Watchung Mountains.

The sediments of the Newark Basin tell the dinosaur story on the middle East Coast from the beginning of the age of dinosaurs well into the Jurassic Period (chapters 5 and 6).

Canada's Fundy

Above the Connecticut Valley, the next exposure of the Newark Supergroup is hundreds of miles to the northeast, in the Fundy Basin of Nova Scotia. The Bay of Fundy's 15 meter (50 foot) tides, the highest on Earth, have exposed dinosaur fossil–bearing sediments of Late Triassic through Early Jurassic age.

The deposits of the Fundy Basin in Nova Scotia are separated into five formations, from oldest to youngest: the Wolfville Formation, Blomidon Formation, North Mountain Basalt, and two equivalent units, the Scots Bay Formation and McCoy Brook Formation. The North Mountain Basalt is the dividing line between Triassic and Jurassic sediments.

Southern Basins

Many of the major rift basins south of the Newark Basin have produced skeletal remains and footprints of reptiles, although not in the abundance or variety of the Newark and the Connecticut Valley basins. These southern basins in Pennsylvania, Maryland, Virginia, and North and South Carolina have historically been studied less than their northern counterparts, perhaps because of their relatively sparse fossil record or less extensive outcrops. The Gettysburg and Culpeper basins expose both Upper Triassic and Lower Jurassic rocks; the remaining southern basins are strictly Upper Triassic.

These southernmost basins apparently lost their Lower Jurassic sediments through erosion, or their sedimentation was completed before the Jurassic began.

Gettysburg Basin

This small basin, exposed in southeastern Pennsylvania and west-central Maryland, is connected by a narrow neck to the western Newark Basin. It also appears to have been connected at one time to the Culpeper Basin to the south. Named for Gettysburg, Pennsylvania, the basin is approximately 130 kilometers (81 miles) long. It extends from Frederick County, Maryland, northeast through Adams, York, and Dauphin counties in Pennsylvania. The Gettysburg Basin has two dinosaur fossil–bearing formations: the New Oxford Formation (the oldest) and the Gettysburg Formation, also known as the Gettysburg Shale. Near the top of the Gettysburg Formation is the Lower Jurassic Aspers Basalt.

Culpeper Basin

The Culpeper Basin, named for Culpeper, Virginia, extends for about 185 kilometers (115 miles) from near the Albemarle County/Orange County line in Virginia northeast to Frederick, Maryland. For nearly 30 million years, from the Late Triassic into the Early Jurassic, it was the site of a huge lake that expanded and shrank with climate changes. At its peak, Lake Culpeper was an estimated 96 kilometers (60 miles) long and 16 kilometers (10 miles) wide.

The geology of the Culpeper Basin is complex, with up to twenty identifiable sedimentary and basalt units. The major formations, several of which have yielded dinosaur tracks, are (from oldest to youngest): the Manassas Sandstone, Balls Bluff Siltstone, Catharpin Creek Formation, Mount Zion Church Basalt, Midland Formation, Hickory Grove Basalt, Turkey Run Formation, Sander Basalt, and Waterfall Formation. The sediments dip to the west and consist largely of red sandstones, shales, and conglomerates.

Richmond and Taylorsville Basins

These two Triassic basins, named for the city and town in Virginia, contain the oldest exposed sedimentary rocks of the Newark Supergroup south of Nova Scotia. The Richmond Basin is approximately 53 kilometers (33 miles) long and is divided into two major stratigraphic units known as the Tuckahoe (oldest) and Chesterfield groups. It extends through the Virginia counties of

Amelia, Chesterfield, Powhatan, Henrico, and Goochland. The Taylorsville Basin—19 kilometers (12 miles) long in Hanover and Caroline counties, Virginia—consists of the Doswell Formation.

Although no dinosaur fossils have been found in either basin, a site southwest of Richmond has yielded the jawbones and other skeletal parts of eucynodonts, very closely related to the earliest mammals. And fishes and primitive archosaurs have been found in the Doswell Formation near the town of Doswell in Hanover County (chapter 9).

Dan River/Danville Basin

There are three formations in the Dan River/Danville Basin, from oldest to youngest: Pine Hall Formation, Cow Branch Formation, and Stoneville Formation. Where these are not differentiated, they have been designated the Dry Fork Formation. These Triassic sedimentary rocks extend from North Carolina into Virginia, with a total length of approximately 175 kilometers (108 miles). They were named for the Dan River in North Carolina and Virginia and for the town of Danville, Virginia. Dinosaur footprints—perhaps

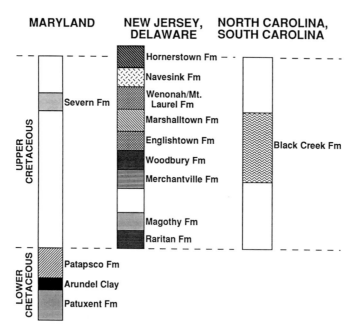

FIG. 3.4 Correlation of Lower Cretaceous–Upper Cretaceous geologic formations, East Coast of North America.

the oldest on the East Coast—have been found in the Cow Branch Formation, a thick series of gray and black siltstones and sandstones.

Deep River Basin

Three Upper Triassic formations or their equivalents are included in the Chatham Group of the Deep River Basin in North and South Carolina, from oldest to youngest: the Pekin Formation, Cumnock Formation, and Sanford Formation. The Deep River Basin—named for the Deep River coalfield in North Carolina—is actually a collection of smaller basins extending about 240 kilometers (140 miles) from Chesterfield County, South Carolina, to Granville County, North Carolina. The Pekin Formation has yielded the only Late Triassic dinosaur fossils from the Deep River Basin.

Jurassic Gap

No dinosaur fossils have been found on the East Coast from Middle Jurassic until Early Cretaceous times, a barren interval of 50–70 million years. Those sediments are missing entirely through erosion or are beyond exploration far offshore to the east. Paleontologists have had better luck looking for Late Jurassic dinosaur fossils in eastern Africa, southern England, and in the Western Interior of the United States (chapter 6).

The long erosion of our Late Jurassic landscape set the stage for the next major act in the drama of East Coast dinosaurs: creation of the Atlantic Coastal Plain (fig. 3.4).

Early Cretaceous Coastal Plain

The dinosaur fossil record in the East resumes in the Early Cretaceous Period. As the Atlantic Ocean opened during the ongoing breakup of Pangaea, the new continental margin of eastern North America subsided. The Appalachian Mountains had been thrust up to spectacular heights to the west during the late Paleozoic, and their subsequent erosion over millions of years built up the coastal plain and continental shelf with layer upon layer of sand, gravel, and silt. Today's Atlantic Coastal Plain ranges in width from 24 kilometers (15 miles) in the north to more than 400 kilometers (250 miles) in the southeastern states.

The western border of the flat, gently sloping plain abuts the rocks of the

Piedmont province at the "fall line," marked by the settlement of major eastern cities like Philadelphia, Pennsylvania; Baltimore, Maryland; Washington, D.C.; and Richmond, Virginia. Locations for the original towns were chosen to take advantage of abundant waterpower to turn mill wheels and accessibility to ocean-bound river transport.

During Early Cretaceous times, much of the East Coast existed as broad, wet lowland plains cut by rivers, similar to the Mississippi Delta today. The lush vegetation included ferns, conifers, cycads, ginkgoes, club mosses, horsetails, and the first angiosperms, or flowering plants. The earliest of the preserved Lower Cretaceous sediments were deposited by flooding rivers along the inner, or western, edge of the Atlantic Coastal Plain. The beds slope gently toward the sea and contain clays and sands in a rainbow of colors: gray, brown, black, red, yellow, purple, white.

Known as the Potomac Group in Maryland, the Lower Cretaceous sediments are exposed from southern New Jersey through Delaware, Baltimore, Washington, D.C., and Fredericksburg, Virginia, to the James River area south of Richmond. Similar Lower Cretaceous clays crop out sporadically north of New Jersey, including the Martha's Vineyard area of Massachusetts.

Geologists have historically identified three divisions within the Potomac Group, from oldest to youngest: the Patuxent Formation, Arundel Clay, and Patapsco Formation. The Arundel was at one time given formation status, but it is now recognized as an intermittent lignitic clay of variable nature and limited scope. The Arundel apparently formed as local pockets of plant-rich sediments laid down atop the Patuxent Formation in oxbow lakes and swamps along a major river drainage.

Neither the Patuxent nor Patapsco formation as historically defined has produced dinosaur fossils. It is the Arundel, the "blue charcoal clay," that represents the only source of dinosaur material from the Early Cretaceous east of the Mississippi River.

Arundel Iron Swamps

The Arundel Clay outcrops irregularly in a 96 kilometer (60 mile) band from Cecil County, Maryland, southwest to Washington, D.C. It reaches a maximum width of 11 kilometers (7 miles) near Laurel, Maryland, and averages 30 meters (100 feet) thick.

Attention was first drawn to the Arundel by its abundant lumps of iron ore, mingled with blackened, charcoal-like wood and fossil plants. The swampy, stagnant conditions that produced the Arundel some 110 Ma were

FIG. 3.5 Reynolds Ore Bank, Anne Arundel County, Maryland. (From J. T. Singewald Jr., *Report on the Iron Ores of Maryland, with an Account of the Iron Industry,* 1911, plate 21)

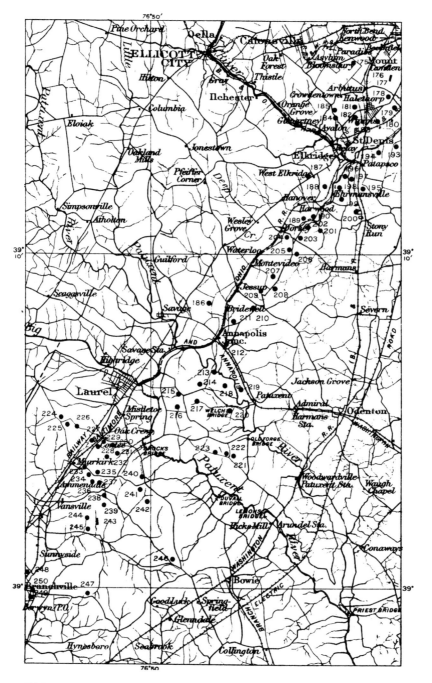

FIG. 3.6 Location of iron ore banks in parts of Baltimore, Howard, Anne Arundel, and Prince Georges counties, Maryland. (From J. T. Singewald Jr., *Report on the Iron Ores of Maryland, with an Account of the Iron Industry,* 1911, plate 24)

apparently ideal for the formation of bog iron, also known as swamp ore or meadow ore. Much of this originally low-grade iron ore was transformed by natural processes into iron carbonate. This form occurs in the Arundel as porous nodules ranging from small chunks to concretions as large as a compact car and weighing several tons.

Beginning in early colonial times and lasting until the World War I period, the iron ore of the Arundel supported a booming mining industry. At its peak value during the Civil War, the iron—prized for its high tensile strength—was made into cannons and cannon balls for the federal army. Arundel iron later found its way into automobile wheels and other modern industrial applications.

The ore was dug by hand in hundreds of surface pits that tapped the narrow Arundel vein from Prince Georges County near Washington, D.C., to north of Baltimore City into Baltimore, Harford, and Cecil counties. The industry was largely centered in two areas of Maryland: Hanover in Anne Arundel County—where the Timber Neck ore banks along Piney Run were extensively worked—and Muirkirk in Prince Georges County (figs. 3.5 and 3.6).

In 1859 an iron pit at Bladensburg in Maryland's Prince Georges County, near Washington, D.C., produced the first dinosaur fossils from the Arundel: two teeth assigned to a new sauropod dinosaur, *Astrodon* (chapter 7). But it is at Muirkirk, and nearby Contee, that the bulk of Early Cretaceous dinosaur fossils have been found (chapters 4 and 9). The most productive dinosaur fossil site still open in the Arundel Clay is near Muirkirk in an active clay quarry operated by Maryland Clay Products. Evidence for additional Arundel dinosaurs was found in Washington, D.C., in 1898 and 1942.

Late Cretaceous Paradise

By the Late Cretaceous, North America resembled its present shape, and the East Coast was relatively stable geologically. Ocean levels were high, with as much as half of the continent underwater. A vast inland sea covered the Western Interior from the Arctic to the Gulf of Mexico, a 4,000 kilometer (2,500 mile) barrier separating the two coasts and their dinosaur populations. Much of the Atlantic Coastal Plain was flooded, as was the lower Mississippi Valley. Although Atlantic waters lapped at the feet of the Piedmont Plateau in some areas, large sections of the gently sloping coastal plain were above sea level and accessible to the Late Cretaceous fauna, including dinosaurs.

The climate was apparently seasonless, with warm temperatures and moderate rainfall spread evenly throughout the year. Flowering plants were well established, and broad-leaved forests and shrubs left their fossil evidence in abundance in Upper Cretaceous sediments. In the widespread seas, marine life was teeming. Giant marine lizards such as the mosasaurs *Tylosaurus* and *Platecarpus* thrived off the East Coast, as did the long-necked *Elasmosaurus*. Sharing the seas with these creatures was *Archelon,* the largest sea turtle that ever lived.

This was the fertile world in which the Late Cretaceous dinosaurs of the East Coast lived. It was an environment dominated by the sea, and dinosaur remains have been found in sediments laid down in coastal waters covering parts of the present states of New Jersey, Delaware, Maryland, and North and South Carolina. During the Late Cretaceous, ocean levels rose and fell at least three times, and the nature of the sediments left behind varies from continental deltaic deposits to fully marine, glauconitic sands.

Nearly a dozen Upper Cretaceous geologic formations have been identified in the Middle Atlantic region, with their greatest development in New Jersey. They are, from oldest to youngest: the Raritan Formation, Magothy Formation, Matawan Group (Merchantville, Woodbury, Englishtown, Marshalltown, and Wenonah formations), and Monmouth Group (Mount Laurel, Navesink, Red Bank, and Tinton formations). Another formation, the Hornerstown, overlies the Monmouth Group and straddles the boundary between the Cretaceous and Tertiary periods.

The Upper Cretaceous formations of New Jersey and Delaware have yielded most of the East Coast dinosaur fossils from this geologic time period. Since all but a few of the fossils were found in marine sediments, paleontologists speculate that the dinosaur carcasses and bones were probably washed into coastal waters from more inland habitats to the west in Pennsylvania, Maryland, and northern New Jersey.

In Maryland, the Upper Cretaceous sediments are grouped into a single marine unit known as the Severn Formation. The Severn has yielded the only Late Cretaceous dinosaur fossils in the state, from exposures in Prince Georges County east of Washington, D.C. (chapter 9).

The Late Cretaceous dinosaur trail picks up again in North Carolina. Here the Black Creek Formation—now properly known in both Carolinas as the Donoho Creek Formation of the Black Creek Group—is exposed in the southeastern part of the state. This formation of sands, silts, and clays—deposited about 75 Ma—correlates with the Matawan Group of New Jersey and Delaware. The prime Black Creek site is at Phoebus Landing on the Cape Fear River, which has yielded most of the dinosaur remains from North Carolina (chapter 9).

Although geologists have known for well over a century that the Upper Cretaceous sediments continue into South Carolina, no dinosaur fossils were found in that state until recently. Several sites in the Donoho Creek Formation in northeastern South Carolina have now yielded dinosaur remains (chapter 9).

Late Cretaceous Greensand Seas

The trademark sediments of the Late Cretaceous on the East Coast are the greensand marls. A "marl" is a sand, silt, or clay with a high calcium carbonate content from fossil shells. "Greensand" marls, or greensands, are rich in glauconite, a complex, bluish green mineral made up of iron, potassium, and other elements. Greensands formed from dead marine plants and invertebrates on the bottom of the warm seas that advanced and retreated over the coastal plain.

Beginning in the late 1700s, greensands found widespread use as an agri-

FIG. 3.7 New Jersey Marl and Transportation Company marl pit at Sewell, New Jersey, in operation about 1930s. The figure in upper left is John C. Vorhees, general manager. (Inversand Company)

cultural fertilizer. Hundreds of private and commercial marl pits were opened in New Jersey during the next century. Just as quarrying for brownstone in the Connecticut Valley and iron ore in Maryland uncovered dinosaur fossils as a by-product of economic activity, the New Jersey marl pits opened a window on the dinosaurs of the Late Cretaceous on the East Coast.

Among the most famous of the fossil-producing sites in New Jersey were marl pits at Barnsboro (West Jersey Marl Company), Hornerstown (Cream Ridge Marl Company), Keyport (Lorillard Company), Marlboro, Swedesboro, and Mullica Hill. One of the most productive of the greensand sites for dinosaur hunters remains a commercial marl pit near Sewell in Gloucester County (fig. 3.7).

The window began to close in the early twentieth century. After reaching a peak in the mid to late 1800s, the marl mining industry rapidly declined with the advent of chemical fertilizers. There was a modest rebirth of the industry when greensand was found to have water-softening properties. As late as the 1960s, there were still three marl operations in New Jersey supplying greensand to a worldwide market for water-softening zeolite. That market thrived until the development of synthetic ion-exchange resins, which proved more efficient than greensand.

Today only the Inversand Company at Sewell is still mining the marls, serving a niche market for "manganese" greensand to remove soluble iron and manganese from domestic, municipal, and industrial water supplies. There also remains a small, steady demand by organic gardeners for greensand as a natural soil conditioner.

4

T H E   P I O N E E R S

Lore and Legend

Native Americans, so in tune with Earth and sky and water, surely noticed the giant bones weathering from the ground and the birdlike footprints preserved on slabs of stone. But their discoveries are lost to modern science; only their legends survive. In North America and many other parts of the world—including Scandinavia, England, and China—folktales abound about dragons, fanged and fire breathing, that ruled the earth in the distant past and left their giant bones for the marvel of mankind.

The Sioux tell of huge serpents that were struck by lightning and burrowed into the earth to die. Other Indian legends describe giant creatures that threatened the ancient fathers of men and angered the Creator with their evil ways. These tales apparently were inspired by fossil remains of the extinct American mammoth, or mastodon, which had been eroding from rocks in North America for thousands of years. In the early 1700s, mastodon bones and teeth were found at such sites as Big Bone Lick along the Ohio River in Kentucky and near the Hudson River in New York State.

A Shawnee legend surrounding five mastodon skeletons from Big Bone Lick was described in 1762 by naturalist James Wright of Pennsylvania:

They had indeed a tradition, such mighty Creatures, once frequented those Savannahs, that there were then men of a size proportionable to them, who used to kill them, and tye them in Their Noppusses And throw them upon their Backs As an Indian now dos a Deer, that they had seen Marks in rocks, which tradition said, were made by these Great & Strong Men, when they sate down with their Burthens, such as a Man makes by sitting down on the Snow, that when there were no more of these strong Men left alive, God Kiled these Mighty Creatures, that they should not hurt the Present race of Indians, And added, God had Kill'd these last 5 they had been questioned about, which the Interpreter said was to be understood, they supposed them to have been Killd by lightning. (James Wright to John Bartram, August 22, 1762, British Museum, Add. MSS 21648, fols. 333–334)

Thomas Jefferson was greatly interested in many of the early fossil finds and is sometimes called the father of American vertebrate paleontology. As governor of Virginia in 1785, Jefferson wrote about a legend told to him by a delegation of warriors from the Delaware tribe:

In ancient times, a herd of these tremendous animals came to the Bigbone licks, and began an universal destruction of the bear, deer, elks, buffaloes, and other animals, which had been created for the use of the Indians; that the Great Man above, looking down and seeing this, was so enraged that he seized his lightning, descended on the earth, seated himself on a neighboring mountain, on a rock, of which his seat and the print of his feet are still to be seen, and hurled his bolts among them till the whole were slaughtered, except the big bull, who presenting his forehead to the shafts, shook them off as they fell; but missing one at length, it wounded him in the side; whereon, springing round, he bounded over the Ohio, over the Wabash, the Illinois, and finally over the great lakes, where he is living at this day.

Noah's Flood

Jefferson and other scholars of his day didn't support the theory of divine lightning bolts, but there was great debate about the ancient animal remains from such sites as Big Bone Lick. Many learned men held theories that by today's standards seem nearly as unscientific as the legends.

Mainstream thought in Western cultures at the time was still guided by the biblical tradition. Archbishop James Ussher of Ireland had declared in 1650 that the earth was created in 4004 B.C. Fossil clues to Earth's antiquity were explained by "Noah's flood," the great cleansing cataclysm of the Bible. Ussher placed the onset of the flood 1,655 years after creation, in December 2349 B.C. Many religious scholars and scientists believed at the time that Earth's layers of rock had formed from accumulations of mud and clay deposited during the great deluge.

Cotton Mather, the New England pastor, naturalist, and "authority" on witchcraft, proclaimed in 1706 that the fossil bones and teeth of a mastodon found along the Hudson River in New York State about 1705 belonged to a race of vanished humans, "godless giants drowned in Noah's Flood."

The idea that the fossils might represent extinct animal species was nearly a century away from popular acceptance, although it was suggested as early as 1768 by William Hunter in London and supported in 1800 by the great French anatomist Baron Georges Cuvier. Charles Darwin's theory of evolution didn't appear in print until 1859, in his *On the Origin of Species by Means of Natural Selection.*

Thomas Jefferson backed the majority view when he openly doubted that extinction could be part of Nature's grand plan. "Such is the economy of nature, that no instance can be produced, of her having permitted any one race of her animals to become extinct; of her having formed any link in her great work so weak as to be broken," Jefferson wrote in 1781. "For if one link in nature's chain might be lost, another and another might be lost, till this whole system of things should evanish by piecemeal."

The First Dinosaurs

The world's earliest recorded dinosaur bones and footprints came to light in the early 1800s, when huge birds and vanished races of giant humans were still acceptable explanations for their troubling existence. It wasn't until 1842 that Sir Richard Owen first recognized Dinosauria (chapter 1).

Owen based his pioneering work largely on three early fossil discoveries made in England. The first fossil bones in history to be identified as a dinosaur belonged to *Megalosaurus,* found in 1818 in Oxfordshire and described in 1824 by the Reverend William Buckland, a geology professor at Oxford University, member of the Royal Society of London and president of the Geological Society of London (fig. 4.1). Discovered in Middle Jurassic deposits in a quarry at Stonesfield, the remains consisted of vertebrae, several hind-

FIG. 4.1 Dean William Buckland described the remains of the world's first dinosaur, *Megalosaurus,* from Stonesfield, England, in 1824.

FIG. 4.2 Mary Ann Mantell found fossils of the world's second known dinosaur, *Iguanodon,* in Sussex, England, in 1822. (Natural History Museum, London)

FIG. 4.3 Gideon Algernon Mantell, a physician and paleontologist, collected *Iguanodon* and described it in 1825. (Natural History Museum, London)

limb bones, and parts of a lower jaw, pelvis, and shoulder blade. Buckland wrote in 1824: "From these dimensions as compared with the ordinary standard of the lizard family, a length exceeding 40 feet and a bulk equal to that of an elephant seven feet high have been assigned by Cuvier to the individual to which this bone [the thighbone or femur] belonged."

Megalosaurus also represented the first meat-eating, or carnivorous, dinosaur ever found, and its description stirred the public imagination. Charles Dickens included a reference in his 1852 novel *Bleak House*: "Implacable November weather. As much mud in the streets as if the water had but newly retired from the face of the earth, and it would not be wonderful to meet a *Megalosaurus,* forty feet long or so, waddling like an elephantine lizard up Holborn Hill."

The second known dinosaur was *Iguanodon,* also from England. In 1822 Mary Ann Mantell (fig. 4.2) reportedly found fossil teeth and bones in a Lower Cretaceous exposure while accompanying her physician husband on his rounds in the Sussex countryside. Gideon A. Mantell (fig. 4.3) was a serious and respected collector of rocks and fossils who was then completing the manuscript of *The Fossils of the South Downs.* Although Dr. Mantell apparently thought from the beginning that the discovery represented a giant reptile, the fossils were misidentified by specialists as either a fish, a rhinoceros, or a hippopotamus. Mantell wasn't convinced, and in 1825 he published "Notice on the *Iguanodon,* a Newly Discovered Fossil Reptile from the Sandstone of Tilgate Forest, in Sussex" in the prestigious *Philosophical Transactions of the Royal Society of London.* He chose the name *Iguanodon* ("iguana tooth") because the fossil teeth resembled those of a modern iguana. We know today that the remains were indeed those of an extinct giant reptile, an herbivorous dinosaur related to the hadrosaurids, or duckbills.

The third fossil discovery that was studied extensively by Owen before his 1842 description of Dinosauria was *Hylaeosaurus,* found in the Wealden of southern England and described by Mantell in 1832. This plant-eating armored reptile is now considered an ankylosaur (chapter 8).

In 1854 Owen and British wildlife artist Benjamin Waterhouse Hawkins collaborated to create life-sized concrete sculptures of *Megalosaurus* and *Iguanodon* for the grounds of London's new Crystal Palace, built for the International Exposition of 1851. Based on the incomplete remains then available, the rhinoceros-like models bear little resemblance to today's more accurate conceptions. In fact *Iguanodon* had a thumb spike mistakenly placed

FIG. 4.4 Sculptures of *Iguanodon* created by Benjamin Waterhouse Hawkins in 1854, still on display at Crystal Palace Park in London.

on its nose. But the exhibits were hugely popular at the time and are still on display in the park (fig. 4.4).

A much earlier fossil discovery in England may also have been a dinosaur. In 1676 Robert Plot, a clergyman and director of the Ashmolean Museum at Oxford University, illustrated part of a large fossil bone that he thought belonged to a giant human. Nearly a century later, in 1763, Dr. Richard Brookes named the fossil *Scrotum humanum* because of its resemblance to giant human testicles. Unfortunately, Plot's bone disappeared long ago and is no longer available for study. Modern researchers note the drawing's similarity to the thighbone of a theropod dinosaur, perhaps *Megalosaurus.*

Another early discovery in England came in 1809, when William Smith was conducting a geological survey of the British Isles. Smith found three large bone fragments at Cuckfield in Sussex, including part of a huge shinbone, or tibia. Many years later the fossils—now in the collection of the Natural History Museum in London—were identified as belonging to *Iguanodon.*

Early American Bones

Across the Atlantic in North America, dinosaur fossils were found years before the discovery of *Megalosaurus* and *Iguanodon.* But the early American finds were misidentified or unrecorded, and their significance wasn't realized until after the bones from England were described.

A "large thighbone" presented before the American Philosophical Society in Philadelphia on October 5, 1787, is now thought to have been a dinosaur foot bone called a metatarsal. Recovered from Upper Cretaceous deposits near Woodbury Creek in Gloucester County, New Jersey, the fossil bone may have belonged to one of the duck-billed dinosaurs, whose remains were later found in those same coastal plain sediments. According to the *Proceedings* of the Society, the bone was introduced by Dr. Caspar Wistar, a prominent Philadelphia anatomist and early paleontologist, and fellow Society member Timothy Matlack. They had no clue to its identity and published no scientific description. The 1787 specimen appears to have vanished soon after, although paleontologist Donald Baird (chapter 9) may have traced it to the collections of the Academy of Natural Sciences in Philadelphia.

In 1806 it was probably a dinosaur bone that caught the eye of explorer William Clark by the Yellowstone River near present-day Billings, Montana. This area of the Upper Cretaceous Hell Creek Formation has since produced many dinosaur fossils. Clark, of the famous Lewis and Clark expedition, described in his journal on July 25, 1806, what he thought were the remains of a large fish. His idiosyncratic spelling and punctuation are faithfully reproduced:

> Dureing the time the men were getting the two big horns [sheep, shot for food] which I had killed to the river I employed my self in getting pieces of the rib of a fish which was Semented within the face of the rock this rib is (about 3) inches in Secumpherence about the middle it is 3 feet in length tho a part of the end appears to have been broken off (the fallen rock is near the water—the face of the rock where rib is is perpend[icula]r—4 i[nche]s lengthwise, a little barb projects I have several pieces of this rib the bone is neither decayed nor petrified but very rotten. the part which I could not get out may be seen, it is about 6 or 7 Miles below Pompys Tower in the face of the Lar[boar]d [north] Clift about 20 feet above the water.

This colorful description is all we know of the Clark discovery, since the bone was never collected and never relocated.

The earliest verified skeletal remains of a North American dinosaur are those found in 1818 in red sandstone blasted from a well in East Windsor, Connecticut. Discovered by Solomon Ellsworth Jr., they were reported in 1820 by Nathan Smith in the *American Journal of Science:*

> Mr. Solomon Ellsworth, Jun., of East Windsor (Conn.), has politely favoured me with some specimens of fossil bones, included in red sand

stone. Mr. Ellsworth informs me that they were discovered by blasting in a rock for a well; they were 23 feet below the surface of the earth, and 18 feet below the top of the rock. Unfortunately, before Mr. Ellsworth came to the knowledge of what was going on, the skeleton had been blown to pieces, with the rock which contained it, and several pieces of bones had been picked up, and then lost. Mr. Ellsworth states that the bones were found in a horizontal position across the bottom of the well, as he thinks nearly to the extent of six feet.

Smith thought they were human remains, although others noted the presence of tailbones. The fossils, now housed in Yale University's Peabody Museum of Natural History, were identified in 1915 as belonging to a prosauropod dinosaur, *Anchisaurus colurus* (chapter 6).

Footprints in Stone

The earliest records of fossil footprints in North America come from the Connecticut Valley, one of the richest sources of dinosaur tracks in the world. Tens of thousands of dinosaur footprints have been found in the Upper Triassic and Lower Jurassic sediments of Massachusetts and Connecticut since 1802. In spring of that year young Pliny Moody was plowing a field on his father's farm in South Hadley, Massachusetts, when he turned up a slab of red sandstone imprinted with mysterious tracks (fig. 4.5). By Moody's account, the 30 centimeter (12 inch) long prints were "three-toed like a bird's," and he proudly installed the slab as a doorstep.

The tracks were said by religious authorities to belong to "Noah's raven," a bird of truly biblical proportions that apparently rested at the South Hadley site before resuming the long journey back to the ark. In fact Pliny Moody's discovery represented the first dinosaur footprints discovered in North America. The prints are now on display at the Pratt Museum of Natural History at Amherst College. Moody was honored in 1847 when the fossil tracks *Otozoum moodi*—now thought to have been made by a large crocodile-like reptile—were named for him.

The next significant footprint discovery in the Connecticut Valley occurred in 1835 at Greenfield, Massachusetts. This find almost single-handedly helped establish the new science of ichnology, the study of fossil tracks. During a town paving project, Greenfield residents noticed "turkey" tracks on sandstone slabs quarried from Turners Falls, Massachusetts, an area that has

FIG. 4.5 Pliny Moody Foot Mark Quarry, South Hadley, Massachusetts, site of first dinosaur footprints discovered in North America in 1802. (From Edward B. Hitchcock, *Ichnology of New England,* 1858)

produced more footprints than any other in the Connecticut Valley. The prints were brought to the attention of a local physician and naturalist, James Deane, who recalled in 1861 that he "recognized in their mute teachings the sublime indications of an Almighty hand [and] entered upon the investigation of the whole subject."

Deane wrote a letter about the discovery to Edward B. Hitchcock, clergyman, professor of natural theology and geology at Amherst College, later the school's president (1844–1854) and state geologist for Massachusetts (1830–1833, 1837–1841) and Vermont (1856–1861). In response to Deane's description and plaster casts of the prints sent to Amherst, Hitchcock (fig. 4.6) rushed to the site. "No facts in my life are more vividly impressed upon my memory than those relating to the footmarks," he wrote in 1858. "As soon as I saw the specimens, I perceived the phenomena worthy of careful research."

Hitchcock had found his life's work. He spent the summer of 1835 criss-

FIG. 4.6 Edward B. Hitchcock of Amherst College pioneered the scientific study of fossil footprints, known as ichnology, in the Connecticut Valley. (Amherst College Archives)

crossing the valley, collecting and studying new specimens. The result was his pioneering 1836 paper, "Ornithichnology: Description of the Footmarks of Birds (Ornithichnites) on New Red Sandstone in Massachusetts." Deane, bitterly overshadowed by Hitchcock's gathering fame, went on to publish his own research on the footprints beginning in 1843 in the *American Journal of Science.*

Hitchcock believed the tracks were made by extinct giant flightless birds, a reasonable conclusion in the context of the early 1800s. It was known at the time that large, ostrichlike moas had recently become extinct in New Zealand, so it wasn't difficult to picture the vanished trackmakers as their huge ancestors. Hitchcock initially described seven footprint species, which he called "ornithichnites," or stony bird tracks. Later, in his 1858 *Ichnology of New England: A Report on the Sandstone of the Connecticut Valley, Especially Its Fossil Footmarks,* he categorized the trackmakers as "pachydactylous, or thick-toed birds," and "leptodactylous, or narrow-toed birds."

Hitchcock wrote:

At first, men supposed that the strange and gigantic races which I had described were mere creatures of imagination, like the Gorgons and Chimeras of the ancient poets. But now that hundreds of their footprints, as fresh and distinct as if yesterday impressed upon the mud, arrest the attention of the sceptic on the ample slabs of our cabinets, he might as reasonably doubt his own corporeal existence as that of these enormous and peculiar races.

Hitchcock left a priceless legacy for later researchers through his careful illustration and studies of both fossil footprints and skeletal remains. He wrote numerous journal articles and a supplement to *Ichnology*, published posthumously in 1865. During his prodigious career, he identified more than 150 species of animals based on their fossil tracks. His extensive collection of prints is today preserved in the Pratt Museum of Natural History at Amherst College in Amherst, Massachusetts.

At the time of his death in 1864, nearly a quarter century after Owen's recognition of Dinosauria, Hitchcock had still not accepted that many of the Connecticut Valley footprints were made by dinosaurs. In fact, today we know that many of the early dinosaurs were remarkably birdlike, and that modern birds descended from a line of reptiles that included dinosaurs (chapter 10).

Hitchcock acknowledged that his footprint interpretations were "tossed on the sea of difficulty." In 1858 he wrote: "That it will be necessary to change the place of some of the species which I have described, I expect. I have done what I could in laying the foundations, and in gathering a storehouse of materials. Let others, with better light to guide them, carry up and complete the structure."

More Early Footprints

Many more dinosaur footprints were discovered in the remaining decades of the nineteenth century and the early twentieth, in the Connecticut Valley, New Jersey, Pennsylvania, Maryland, and Virginia. Detailed scientific study of the tracks slowed considerably after Hitchcock's death, although his son, Charles H. Hitchcock, was himself a distinguished researcher and described a number of new ichnospecies.

It wasn't until 1904 that Richard Swann Lull took up the torch and published his first study of Connecticut Valley footprints, "Fossil Footprints of the Jura-Trias in North America." Lull (fig. 4.7) became a major figure in East Coast paleontology during his long career as director of Yale University's Peabody Museum of Natural History (chapter 9). "One of the most interest-

FIG. 4.7 Richard Swann Lull (in 1952) of Yale University was a major figure in East Coast dinosaur paleontology in the first half of the twentieth century. (Courtesy of the Peabody Museum of Natural History, Yale University)

ing chapters of the earth's past history," Lull wrote, "is that of the time when there were laid down the Triassic strata of the famed Connecticut Valley, interesting in the profusion of its indicated life, and fascinating in the baffling obscurity which shrouds most of its former denizens, the only records of whose existence are 'footprints on the sands of time.'"

From elsewhere on the East Coast, early footprint specimens included those found in 1886 by John Eyerman near Milford, in Hunterdon County, New Jersey. Eyerman reported two footprint-bearing slabs in a brief description, "Footprints on the Triassic Sandstone of New Jersey," in the *American Journal of Science*. These tracks were later referred to *Atreipus*, made by a small, primitive ornithischian dinosaur that walked on all fours and was apparently very common on the East Coast throughout the Late Triassic and Early Jurassic (chapters 5 and 6).

In 1895 the first and only dinosaur footprints ever found in Maryland came to light. James A. Mitchell discovered and sketched tracks in a quarry north of Emmitsburg, in Frederick County in the west-central part of the state.

Mitchell was a forty-three-year-old professor of geology, mathematics, and natural science at Mount St. Mary's College in Emmitsburg and a graduate student in geology at the Johns Hopkins University in Baltimore. The tracks, found in the Upper Triassic Gettysburg Formation, were three-toed, from 6.5 centimeters (2.5 inch) to 9 centimeters (3.5 inches) long with a 30 centimeter (12 inch) stride between them. They represent the most common dinosaur footprint on the East Coast, now known as *Grallator* (chapter 5).

Pennsylvania also yielded early dinosaur tracks. Among those recorded are discoveries in 1889 at a small quarry near Goldsboro in York County and in 1902 at Fisher's Quarry near Graterford in Montgomery County. Both locations yielded *Atreipus* footprints.

South of Pennsylvania and Maryland, in the rolling horse country of Loudoun County, Virginia, dinosaur footprints were found near Aldie at Oak Hill, the home of President James Monroe from 1808 to 1831. The stately brick mansion (fig. 4.8), now a National Historic Landmark and a Virginia Historic Landmark, underwent extensive renovation about 1920. As part of this work, its walkways and beautiful terraced gardens were paved with flagstones quarried from a Lower Jurassic outcrop of the Culpeper Basin on the northern border of the huge estate, about a mile from the main house.

When workmen began laying the slabs, they noticed raindrop impressions, ripple marks, and birdlike, three-toed footprints in the flagstones. The dinosaur tracks, known as *Grallator* and *Eubrontes* (chapter 6), range in length from about 13 to 33 centimeters (5 to 13 inches). The distinctive footprint-bearing slabs are still in place today, and a trackway of four prints is actually inside the mansion, on the flagstone floor of an enclosed side porch (figs. 4.9 and 4.10). The inside slabs are a burnished amber hue from decades of diligent waxing by Oak Hill housekeepers. The estate is now owned by Thomas DeLashmutt and family.

Back in New Jersey, the only Cretaceous Period dinosaur footprints ever found in eastern North America came to light in 1929 and 1930. The 90 million year old tracks were discovered in the Hampton Cutter Clay Works pit at Woodbridge, in Middlesex County, near Perth Amboy. Similar tracks had turned up in the same quarry fifteen years earlier but were destroyed during an attempt to remove them. The new discoveries included several trackways of up to four prints each of a clawed, three-toed foot with a maximum length of 48 centimeters (19 inches). Although the prints were removed to the Rutgers University Geological Museum, only a single specimen reportedly survives there. It represents an indeterminate large meat-eating dinosaur.

In the 1930s, a rash of dinosaur footprint discoveries in southeastern Penn-

FIG. 4.8 Oak Hill mansion near Aldie, Virginia, home of President James Monroe from 1808 to 1831.

sylvania helped expand knowledge of that state's dinosaur heritage. Two tracks were found in 1933 near Yocumtown in York County and identified as *Anchisauripus*. In 1933 and 1937, Late Triassic dinosaur footprints were discovered near the Civil War battlefields in Gettysburg. The latter find included tracks described in a news account as "about six inches long, with a stride of about thirty inches; other, smaller tracks indicate that some of the dinosaurs were no bigger than chickens."

Another Pennsylvania discovery occurred in 1934 near New Cumberland, just southwest of Harrisburg across the Susquehanna River, during widening of Route 111. A single dinosaur track was found on a slab of red shale, along with mud cracks, raindrop impressions, and the prints of plant stems. Bradford Willard of the Pennsylvania Topographic and Geologic Survey was credited with the find.

Also in 1937, tracks were collected by Elmer R. Haile Jr. in the Trostle Quarry near York Springs in Adams County. Among the Late Triassic footprints discovered were *Atreipus*, plus the primitive archosaurian reptile tracks known as *Brachychirotherium* and *Rhynchosauroides* (chapter 5).

An important early figure in Pennsylvania paleontology was Earl L. Poole, who served from 1926 to 1957 as assistant director and director of the Public Museum and Art Gallery in Reading. Poole (fig. 4.11) was a knowledgeable fossil collector as well as a widely known wildlife artist and sculptor. In early 1939 he discovered dozens of reptile footprints in Upper Triassic sediments along Perkiomen Creek near Schwenksville, Montgomery County, about 40 kilometers (25 miles) from Reading. Highway workers had cut into a hillside to quarry rock for roadwork, at a site known as the Squirrel Hill Quarry.

Tracks of three sizes were found, according to a newspaper account: "One

FIG. 4.9 Enclosed porch at Oak Hill mansion, floored with paving slabs imprinted with dinosaur tracks.

FIG. 4.10 Inset from Oak Hill porch showing large three-toed dinosaur footprint in floor slab.

FIG. 4.11 Earl L. Poole, longtime director of the Reading Public Museum, discovered dinosaur tracks in Pennsylvania in the 1930s. (Historical Society of Berks County)

very small, made by a dinosaur about the size of a chicken; one apparently made by a dinosaur about the size of a turkey; one made by an animal whose weight and bulk would approximate that of a horse. There are some scores of the small prints, half dozen of the medium-sized, only one of the large." Poole identified the medium-sized tracks as *Anchisauripus*. In 1952 Wilhelm Bock (chapter 9) referred the Squirrel Hill prints to *Grallator* and several primitive, nondinosaurian reptiles.

Another source of Pennsylvania footprints was a spectacular Upper Triassic exposure along the Reading Railroad tracks at Gwynedd (fig. 4.12). Originally constructed in the mid-1800s as a tunnel for the North Pennsylvania Railroad, it was enlarged to an open cut in 1923. The long, steep cut has yielded reptile prints and skeletal remains plus abundant fish, invertebrate, and plant fossils. In 1952 Bock identified a dinosaur track from Gwynedd as *Anchisauripus gwyneddensis,* now referred to *Atreipus*.

Finally, the South Hadley, Massachusetts, vicinity of Pliny Moody's 1802 discovery of North America's first dinosaur tracks began yielding additional prints in the 1930s. Carlton S. Nash—who grew up within sight of the Moody Quarry—found a slab bearing dinosaur footprints in 1933 on a nearby two-acre outcrop of the Lower Jurassic Portland Formation. Six years later Nash bought the property. He has since found—and sold to collectors from around the world—thousands of reptile prints from his still-producing quarry, now known as Nash Dinosaurland. Many of the prints were made by the same

FIG. 4.12 The Gwynedd Railroad cut in southeastern Pennsylvania, a historic East Coast dinosaur fossil site, shown in 1993.

FIG. 4.13 Carlton Nash found his first dinosaur tracks in 1933 near Pliny Moody's farm in South Hadley, Massachusetts. (Courtesy of Carlton Nash, Nash Dinosaurland)

theropod dinosaur trackmakers that left their mark throughout the Connecticut Valley (fig. 4.13).

Haddonfield *Hadrosaurus*

Many observers date the real beginning of North American dinosaur paleontology to the description in 1858 of the world's first reasonably complete dinosaur skeleton. For the first time in history, enough of a skeleton had been recovered to accurately reconstruct the nearly full appearance of a dinosaur.

The story had begun twenty years earlier, on the farm of John E. Hopkins in Haddonfield, New Jersey, just east of Philadelphia across the Delaware River. Workers excavating marl for fertilizer hit a cache of ancient bones on the Hopkins property, along a tributary of the Cooper River. The fossils lay in Upper Cretaceous marine sediments, and most were huge vertebrae. Hopkins remembered that no skull, leg bones, or other large skeletal parts had been found except one that resembled a "shoulder blade." Unfortunately, as was later reported, "Mr. Hopkins, being young at the time of the discovery, and not specially interested in such subjects, had permitted visitors to carry away the fossils, so that none remained in his possession, nor could he remember the names of any of the persons by whom the vertebrae had been taken." Rumors persisted for years that the souvenir vertebrae had found a new life in southern New Jersey homes as window props and doorstops.

In 1858 William Parker Foulke heard about the bones while summering in Haddonfield. Foulke, a gentleman naturalist and a member of the Academy of Natural Sciences of Philadelphia, reasoned that more bones might be found in the filled and overgrown marl pit. Based on recollections of its location by Hopkins and the original diggers, Foulke hired a crew to reopen the pit that fall. Approximately 10 feet down, they found a 75 million year old dinosaur fossil treasure, described in 1858 by Joseph Leidy of the University of Pennsylvania: "The bones consist of 28 vertebrae, mostly with their processes lost; a humerus, radius, and an ulna complete; an ilium and a supposed pubic bone, imperfect; a femur and tibia complete; a fibula, with one end lost; two metatarsal bones and a phalanx, complete; two small fragments of jaws, and nine teeth."

Leidy (fig. 4.14)—a medical doctor and professor of anatomy who was one of the great pioneers in dinosaur paleontology—had two years earlier officially named the first dinosaurs in North America from some teeth found in

FIG. 4.14 Joseph Leidy (in 1863), founder of the science of vertebrate paleontology in North America, described *Hadrosaurus foulkii* in 1858. (Library, the Academy of Natural Sciences of Philadelphia)

the awakening fossil fields of the western United States. At Foulke's invitation, Leidy traveled to Haddonfield and lent his expertise to the excavation. "The marl being tenacious, great care was requisite to extricate the fossils. With a small trowel and a knife, the bones were carefully dissected from their bed, and from one another," Leidy wrote in 1858. "A sketch was made of their position, and some measurements were taken of them, in anticipation of the contingency of their fracture in the attempt to remove them. The bones are ebony-black, firm in texture, heavy, and strongly impregnated with [iron] salts. . . . They are generally well-preserved, except that many are fractured, but none are water rolled, and a few specimens only appear somewhat crushed."

Leidy immediately noticed "the great disproportion of size" between the front and back parts of the skeleton. That was all the evidence he needed to describe for the first time a dinosaur that walked on two legs: "[It] leads me to suspect that this great extinct herbivorous lizard may have been in the habit of browsing, sustaining itself, kangaroo-like, in an erect position on its back extremities and tail."

In honor of Foulke, Leidy christened the dinosaur *Hadrosaurus foulkii* in

FIG. 4.15 *Hadrosaurus foulkii* at the Academy of Natural Sciences of Philadelphia in 1868, the first dinosaur skeleton mounted for exhibit. (Library, the Academy of Natural Sciences of Philadelphia)

1858. Leidy further interpreted the fossil clues to describe the beast as "most probably amphibious; and though its remains were obtained from a marine deposit, the rarity of them in the latter leads us to suppose that those in our possession had been carried down the current of a river, upon whose banks the animals lived."

In 1868 *Hadrosaurus foulkii* became the first dinosaur skeleton ever mounted for exhibit (fig. 4.15), at the Academy of Natural Sciences of Philadelphia. Benjamin Waterhouse Hawkins collaborated with Leidy to reassemble the bones in a realistic pose. Since the head and other skeletal parts were missing, these were approximated with plaster reconstructions. The exhibit proved one of the most popular in Academy history, increasing attendance by as much as 50 percent. Hawkins greatly influenced the public image of dinosaurs at the time through his well-known paintings of these extinct giant reptiles and their exotic habitats (fig. 4.16).

More than 115 years later, on September 29, 1984, a small monument to

FIG. 4.16 Benjamin Waterhouse Hawkins's 1877 restoration of Late Cretaceous dinosaurs from New Jersey. (Courtesy of the Department of Geological and Geophysical Sciences, Princeton University)

FIG. 4.17 Monument in Haddonfield, New Jersey, to *Hadrosaurus foulkii,* the first nearly complete dinosaur skeleton to be discovered.

Hadrosaurus foulkii was dedicated near the Haddonfield site (fig. 4.17). The monument, funded by the Academy, was the Eagle Scout project of Christopher Brees of Haddonfield. *Hadrosaurus foulkii* was subsequently designated the official "New Jersey State Dinosaur" in a bill signed by then-governor James Florio on June 13, 1991. The bill reads in part: "This discovery . . . was so unexpected and unusual that it startled the scientific thinking of the day and led to a revision of many conventional ideas as to the physical structure and life habits of prehistoric reptiles and provided a great stimulus to the study of dinosaurs which, until then, were relatively unknown outside the scientific community."

The East Coast Awakes

The announcement of *Hadrosaurus foulkii* in 1858 was a pivotal event in dinosaur paleontology, but it was far from the only significant early discovery of dinosaur bones on the East Coast. In 1855 a find occurred at Springfield, Massachusetts. As related by Hitchcock, "The Springfield bones were discovered by William Smith, Esq., while engaged in superintending some improvements at the water shops of the United States Armory, which required blasting. He did not discover them till a large part had been taken away by the workmen. General Whitney, superintendent of the armory, very kindly ordered a re-examination of the fragments, and Mr. Smith obligingly presented me with whatever pieces could be found. These I put into the hands of Professor Jeffries Wyman."

The Springfield bones came from Lower Jurassic sediments and included eleven vertebrae, an incomplete scapula, an almost complete hand, part of the pelvis, and an incomplete hind limb. The bones were examined by Sir Richard Owen in England, and in 1865 the specimen was assigned the name *Megadactylus polyzelus*. It is now referred to the prosauropod *Anchisaurus polyzelus* (chapter 6).

Evidence for another new dinosaur species came in 1859 from an iron mining pit in Bladensburg, Prince Georges County, Maryland, near Washington, D.C. Philip T. Tyson, state agricultural chemist for Maryland, discovered two curiously shaped fossil teeth in the Lower Cretaceous Arundel Clay. Tyson showed the dinosaur teeth to Dr. Christopher Johnston, a local physician, who thinly sliced one for microscopic study and thus noted the tooth's "star-like" appearance in cross section. Johnston named the dinosaur *Astrodon* in his "Note upon Odontography" in the *American Journal of Dental Science* in 1859. The Bladensburg teeth were formally described by Leidy in 1865 as *Astrodon johnstoni,* in honor of Dr. Johnston. *Astrodon* was the first sauropod dinosaur to be named and described in North America (chapter 7).

In that same year (1865), Leidy described a single tooth found in Upper Cretaceous greensand deposits at Mullica Hill, New Jersey. Although broken and badly worn, the tooth had the characteristic serrated edges of a carnivorous dinosaur. Leidy named the dinosaur *Diplotomodon horrificus*, now thought to be an indeterminate theropod (chapter 8).

Marsh and Cope

In the 1860s, two brilliant paleontologists entered the picture, archrivals who energized—and sometimes scandalized—the field of dinosaur hunting with their heated competition. By the last years of the nineteenth century, between them they had described and named more than 135 new dinosaur species.

Edward Drinker Cope was a precocious student from a wealthy Quaker family in Philadelphia (fig. 4.18). With an intense, early interest in paleontology, he frequented the Academy of Natural Sciences during his youth and published his first scientific paper in the Academy's *Proceedings* when he was just eighteen years old. Cope became a top anatomy student of Leidy's at the University of Pennsylvania and later studied at the Smithsonian Institution and abroad during the Civil War. He married at age twenty-five and taught

FIG. 4.18 Edward Drinker Cope raced his archrival Othniel Charles Marsh to describe new dinosaur species from throughout North America.

FIG. 4.19 Othniel Charles Marsh, the nation's first professor of paleontology, worked single-mindedly to expand the dinosaur menagerie in the late 1800s. (Courtesy of the Peabody Museum of Natural History, Yale University)

zoology and botany at Haverford College until taking up fieldwork in dinosaur paleontology full time in 1867.

Othniel Charles Marsh was equally intense and fascinated by fossil collecting, though reportedly less brilliant than Cope. Through the generosity of his famous uncle, the banker and philanthropist George Peabody, Marsh (fig. 4.19) graduated from Yale University and studied in Europe. In 1865 Peabody endowed a natural history museum at Yale, with his nephew installed as the first professor of paleontology in the United States. Marsh was later appointed by John Wesley Powell as national paleontologist to the newly created United States Geological Survey. He was a lifelong bachelor, single-mindedly dedicated to his work.

Cope and Marsh were friendly colleagues at first, sharing information about fossil sites and discoveries on the East Coast. Cope was particularly interested in the Upper Cretaceous marl pits of New Jersey and moved his family to a house in Haddonfield to be near them. He regularly visited the pits and arranged with the diggers to send him interesting fossil bones.

In 1866 a partial dinosaur skeleton was discovered in the "chocolate" marl in a pit operated by the West Jersey Marl Company near Barnsboro, Gloucester County, New Jersey. The skeleton included portions of the lower jaw with teeth, two humeri, left femur, tibia, and fibula, numerous vertebrae, and many fragments. Cope named this bipedal carnivorous dinosaur *Laelaps aquilunguis,* a swift and ferocious predator that slashed prey with long, razor-sharp front claws (fig. 4.20).

The *Laelaps* discovery attracted the attention of Marsh, who toured the marl pits as Cope's guest in 1868. In the apparent first shot in their decades-long duel, Marsh used the occasion to quietly make his own contacts among the workers. Soon the bones began detouring to Marsh's laboratory at the Yale Peabody Museum instead of to Cope at the Academy in Philadelphia. To make matters worse between the two men, Marsh publicly pointed out that

FIG. 4.20 Artist Charles R. Knight's 1897 depiction of *Laelaps* (*Dryptosaurus*) in combat. (Courtesy of Department of Library Services, American Museum of Natural History)

the name *Laelaps* was already used, or "preoccupied," by an insect. He renamed the species *Dryptosaurus,* which stands to this day (chapter 8).

These indignities compounded the unpleasantness over Cope's error in switching the head and tail on the mounted skeleton of a long-necked plesiosaur known as *Elasmosaurus* at the Academy. Marsh loudly trumpeted the mistake to the scientific community, and Cope never forgave him. Of course it was Marsh who too hastily attached the wrong head to a *Brontosaurus* skeleton, an error that denied *Apatosaurus* its true body form for nearly a century.

Thus began the great "bone wars" between the two men, which, among other consequences, apparently forced the gentlemanly Joseph Leidy out of vertebrate paleontology in disgust. Cope and Marsh spent the rest of their lives racing to discover and describe new dinosaurs, largely from the vast new fossil fields in the American West, but also from the small quarries and railroad cuts of the East Coast.

More East Coast Bones

In 1869 the Reverend Samuel Lockwood found dinosaur bones in an Upper Cretaceous clay bank along the shore of Raritan Bay at Union, near Keyport, New Jersey. Cope called it a "remarkable fragment of a gigantic Dinosaur" and praised Lockwood's diligence in making the discovery. Based on very fragmentary evidence consisting of the lower bones of the leg and the anklebones, Cope named the Keyport dinosaur *Ornithotarsus immanis.*

Curiously, Lockwood fully intended for his dinosaur fossils to go to Marsh at Yale. But according to a letter from Lockwood to Marsh in May 1869, an aggressive "gentleman" appeared at his door just before the bones were to be shipped. "He had a very large carpet bag and announced himself as Prof. Cope," Lockwood wrote. "He had smelt the bone and come for it, but no sir that could not be. Well he must see it, and see it he did, and took drawings of it. . . . I can say most literally as regards Prof. Ed. C. that this humble individual could not Cope with him. He is the man to seize your very bones!" Marsh eventually acquired the fossils as intended, but they remained an irritant to the end of his days: a new dinosaur in the Yale collection named by his archrival Cope.

In 1896 a single foot bone of this same dinosaur, *Ornithotarsus immanis,* was found by Lewis Woolman in Merchantville, Camden County, New Jersey, during excavation for an underpass on the Pennsylvania Railroad. "The bone is 15 inches long, and near the middle of the shaft is four inches wide and two and one-half inches thick," Woolman wrote. "On submitting the specimen to Prof. E. D. Cope, he [Cope] pronounced it the middle metatarsal of the right hind limb of an animal of the family of dinosaurian reptiles known as the Hadrosaurs." *Ornithotarsus* is now considered a large individual of *Hadrosaurus foulkii* (chapter 8).

In 1870 Marsh named a smaller New Jersey duckbill *Hadrosaurus minor*. A number of its vertebrae were discovered in the West Jersey Marl Company pit at Barnsboro and—as the first fruit of Marsh's backdoor efforts in the marl pits—sent to Yale rather than to Philadelphia. Other specimens of *H. minor* were later found in New Jersey at Mullica Hill and Sewell.

In the early 1900s a sand pit at Roebling, on the south bank of the Delaware River in Burlington County, New Jersey, yielded the broken end of a foot bone of a large carnivorous dinosaur. Most likely from the Upper Cretaceous Raritan Formation (chapter 8), the metatarsal fragment was donated to Princeton University by Dr. Charles C. Abbott of the State Geological Survey and is now in the collection of the Yale Peabody Museum.

Farther south along the East Coast, North Carolina yielded its first dinosaur about 1869, although Ebenezer Emmons had discovered Cretaceous reptile remains before that during his tenure as state geologist from 1851 to 1863. Emmons's successor, Washington Caruthers Kerr, continued the energetic search for fossils in the Upper Cretaceous of North Carolina. Kerr found dinosaur bones in Sampson County in the marl pit of James King, "some 10 miles from the depot [Faison's Depot], southwest. This locality is on the waters of Six Runs Creek." Cope described this dinosaur in 1869 as the duck-billed *Hypsibema crassicauda* (chapter 8).

The only dinosaur skeletal remains ever found in Pennsylvania are two teeth described in 1878 by Cope, who identified their owner as *Thecodontosaurus gibbidens*. Originally the teeth were thought to have come from the Black Rock railroad tunnel at Phoenixville in Chester County, a famous fossil site that at one time was the longest tunnel excavation in North America. Modern researchers place the source instead near Emigsville in York County, in Upper Triassic sediments of the Gettysburg Basin. In 1994 the teeth were renamed *Galtonia gibbidens*, a primitive ornithischian dinosaur (chapter 5).

Original Academy of Natural Sciences building at Nineteenth and Race Streets, Philadelphia. (Library, the Academy of Natural Sciences of Philadelphia)

THE GREAT INSTITUTIONS

Before the opening of the great American West to scientific investigation, a host of eastern institutions encouraged, supported, and published the results of research on the geology and paleontology of the East Coast. The development of vertebrate paleontology went hand in hand with the rise of learned societies, museums of natural history, and research and teaching departments in the discipline at major East Coast universities and colleges. As paleontologist George Gaylord Simpson observed in 1942, "Vertebrate paleontology has been the most important single factor in the rise and popularization of natural history museums. This is perhaps its greatest contribution to the social, as opposed to the scientific, history of America."

Philadelphia was the early center of New World paleontology. The American Philosophical Society (APS), the first learned society in the United States, was founded in that city in 1743 by Benjamin Franklin and James Logan "for the promotion of useful knowledge among the British plantations in America." The APS was a major force behind the successful Lewis and Clark expedition of 1804–1806, which provided the first reliable records of western fossil localities. Scientific publishing in the United States began in 1769 with the Society's *Transactions;* the APS *Proceedings* commenced publication in 1838.

The Academy of Natural Sciences of Philadelphia was founded in 1812 and provided a popular forum for paleontologists in its *Journal,* beginning in 1823, and the *Proceedings* in 1841. Both the Academy and the American Philosophical Society were closely allied with the University of Pennsylvania and served as a research focus for such early luminaries in the field as Caspar Wistar, Joseph Leidy, and Edward Cope. Leidy and Cope were avid workers on the dinosaurs of New Jersey and studied the first fossil vertebrate material (including dinosaurs) discovered in the pioneering surveys of the American West.

The Lyceum of Natural History of New York was another early focus for paleontological research. It was organized in 1817 through the efforts of Samuel L. Mitchill and served as home base for pioneering zoologist James E. DeKay. The *Annals* of the Lyceum began publication in 1823.

In 1818 geologist Benjamin Silliman of Yale University founded the *American Journal of Science,* probably the greatest source of documentation on early dinosaur fossil finds in North America. Yale and its Peabody Museum of Natural History (founded in 1866) boasted dinosaur paleontologist Othniel Charles Marsh. Marsh and his exceptional collector John Bell Hatcher were to amass a great wealth of dinosaurs and other fossil vertebrates at the end of the 1800s. This material is now split between the Yale Peabody Museum and the U.S. National Museum of Natural History. With Marsh's death in 1899, Richard Swann Lull took over curatorial responsibilities at Yale, advancing knowledge on the life of the Connecticut Valley in addition to his other work on dinosaur biology and taxonomy elsewhere in North America.

Amherst College, founded in 1821, provided a home for the dinosaur footprints collected and described by Professor Edward B. Hitchcock. The collection of more than eight thousand tracks was originally displayed in the "Appleton Cabinet," a museum specially built in 1855 for the footprint slabs. The cabinet was funded through a bequest to Amherst in the will of Samuel Appleton of Boston.

The Appleton Cabinet at Amherst College, repository for Edward Hitchcock's collection of Connecticut Valley footprints. (Amherst College Archives)

The U.S. National Museum in Washington, D.C., part of the Smithsonian Institution, was founded in 1846. Among the Smithsonian's famous dinosaur researchers was Charles W. Gilmore, curator of the museum from 1923 to 1945. It was Gilmore who revised much of the Arundel fauna, particularly that collected by Hatcher and Arthur B. Bibbins. The *Smithsonian Contributions to Knowledge* series published many early paleontological discoveries.

The American Museum of Natural History (AMNH) was organized in 1869 in New York. Paleontologist Henry Fairfield Osborn founded the department of vertebrate paleontology and served twenty-five years as AMNH president. During his tenure the AMNH was a major force behind pioneering collecting campaigns in the American West and Canada, and the Gobi Desert in Mongolia. Among the great dinosaur hunters associated with the museum were Barnum Brown, Walter Granger, and Roy Chapman Andrews. The AMNH monthly magazine, *Natural History,* began publication in 1897.

Rounding out the roster of great institutions is the Carnegie Museum in Pittsburgh, endowed in 1895 by industrialist and philanthropist Andrew Carnegie. A Carnegie-supported expedition led by Earl Douglass discovered the rich fossil site in Utah that would become Dinosaur National Monument.

The Manchester Gold Mine

The mother lode for dinosaur remains in the Connecticut Valley was a quarry in Manchester, Connecticut, that produced three skeletons of Early Jurassic prosauropod dinosaurs during a brief period in the late 1800s (chapter 6). Marsh described the site as "the quarry of Mr. Charles O. Wolcott, about one mile north of Buckland station in a layer about two and one-half feet thickness, and, as the quarry was then worked, somewhat above the level of the roadway."

The first skeleton was discovered on October 20, 1884. "The remains of this reptile are from the sandstone of the Connecticut River Valley, which has long been known for the great variety of footprints it contains, especially those supposed to have been made by birds," wrote Marsh, who described the fossils. "The extreme rarity of any bones in these beds is equally well known, not more than half a dozen finds having yet been made, and only a few of these of much scientific interest."

Workmen excavating sandstone from the quarry discovered fossil bones of the hind limbs and pelvis embedded in a large stone block. It appeared to Marsh that the skeleton may have been complete before the cut. But the blocks likely to contain the skull and forelimbs had already been removed and built into the abutments of a bridge in South Manchester, 6.5 kilometers (4 miles) south of the quarry. Marsh swallowed his disappointment and named the dinosaur *Anchisaurus major,* a small prosauropod. It was later re-named *Ammosaurus major* (chapter 6).

The second skeleton recovered from the Wolcott Quarry was described by Marsh in 1892 as "perhaps the most perfect Triassic Dinosaur yet discovered, as the skull and greater portion of the skeleton were found in place, and in fine preservation." Named *Anchisaurus colurus* by Marsh, it is known today as *Anchisaurus polyzelus* (chapter 6). It remains the most complete and best-preserved skeleton of a prosauropod dinosaur ever found in North America.

The last of the three skeletons was found in the quarry at the same time as *Anchisaurus colurus*. This almost complete skeleton is poorly preserved and smaller than the others. It was named *Anchisaurus solus* by Marsh but is now thought to be a juvenile *Ammosaurus major* (chapter 6).

The Dinosaur Lady

The Connecticut Valley "bone rush" ended in 1910 with a small dinosaur skeleton discovered in South Hadley, Massachusetts—not far from Pliny

FIG. 4.21 Mignon Talbot of Mount Holyoke College in South Hadley, Massachusetts, discovered *Podokesaurus holyokensis* in 1910. (Mount Holyoke College Archives)

Moody's farm—by one of the few women in early North American dinosaur paleontology.

Mignon Talbot was professor and chair of geology and geography at Mount Holyoke College, a fellow of the Geological Society of America (1913), and the first woman geologist to be elected to the Paleontological Society (1909). Talbot (fig. 4.21) was a popular and energetic, though physically tiny, teacher who delighted in taking student groups into the field on collecting expeditions.

Mount Holyoke already had a small collection of dinosaur tracks and other fossils when Talbot joined the faculty in 1904 after earning her doctorate from Yale. But she brought the little geological museum in Mount Holyoke's Williston Hall its most prized specimen in October 1910 with her fossil discovery in a gravel pit on the John A. Boynton farm near South Hadley Center. The exposure was in Lower Jurassic sediments, originally considered Upper Triassic.

"In a bowlder of triassic sandstone which the glacier carried two or three miles, possibly, and deposited not far from the site of Mount Holyoke College, the writer recently found an excellently preserved skeleton of a small dinosaur the length of whose body is about 18 cm," Talbot recalled. "The bowlder was split along the plane in which the fossil lies, and part of the bones are in one half and part in the other. These bones are hollow and the whole framework is very light and delicate. As the fossil lies in the rock, most of the bones are in position, or nearly so, with the exception of the skull and the tail. A detached tail that probably belongs to this specimen lies a few centimeters from the rest of the skeleton and near it are three very thin bones that may belong to the skull."

Talbot originally described the skeleton as an herbivorous dinosaur at a meeting of the Paleontological Society in December 1910. But her subsequent investigation, aided by Richard Swann Lull at Yale, identified it as a small theropod dinosaur that she named *Podokesaurus holyokensis* in 1911 (chapter 6). The bones of *Podokesaurus* were tragically lost in a fire that destroyed Williston Hall in 1916, but casts of the skeleton exist in the Peabody Museum at Yale and the American Museum of Natural History in New York.

Hatcher and Bibbins

In the waning years of the nineteenth century, two diligent collectors pioneered exploration in the Arundel Clay of Maryland. Their efforts produced nearly the entire collection of dinosaurs from the Early Cretaceous of the East Coast.

John Bell Hatcher was Marsh's master collector at Yale, a born fossil sleuth who earned these accolades in his 1904 obituary in *Science* magazine: "[Hatcher had] marvelous powers of vision, at once telescopic and microscopic, a dauntless energy and fertility of resource that laughed all obstacles to scorn. . . . He may be said to have fairly revolutionized the methods of collecting vertebrate fossils, a work which before his time had been almost wholly in the hands of untrained and unskilled men, but which he converted into a fine art."

Hatcher (fig. 4.22) made his reputation in the fossil fields of the American West, beginning upon his graduation from Yale in 1884. During his twenty-year collecting career, he unearthed and shipped an estimated 1,500 large crates of fossils back east to the Smithsonian, Carnegie, Peabody, and Princeton museums. In 1889 Hatcher discovered in Wyoming the skull and other skeletal remains of a three-horned dinosaur named by Marsh *Triceratops*, the first of some fifty individual ceratopsians he would recover by 1892.

But Marsh and Hatcher hadn't forgotten the East Coast. Intrigued by the discovery in 1859 of the *Astrodon johnstoni* teeth in Bladensburg, in late 1887 Marsh dispatched Hatcher to scout out the Arundel Clay exposures in Maryland. A booming iron-mining industry had opened numerous small surface pits between Baltimore and Washington in the fossil-bearing sediments (chapter 3), and the diggers were reportedly finding dinosaur bones.

Hatcher soon located what turned out to be the richest dinosaur fossil site

FIG. 4.22 John Bell Hatcher of Yale was Othniel Charles Marsh's master dinosaur collector, including those from East Coast sites. (Courtesy of the Peabody Museum of Natural History, Yale University)

ever found in the Lower Cretaceous of the East Coast, in "a bed of iron ore near [Muirkirk] Maryland. The exact locality was certain iron ore mines on the farm of Mr. William Coffin, and especially in that one locally known as 'Swampoodle' and situated about 1.5 miles northeast of Beltsville, some 13 miles from Washington." On the main line of the Baltimore and Ohio Railroad, Muirkirk was the site of the major regional iron furnaces plus numerous surface pits with colorful names like Swampoodle, Coffin's Old Engine Bank, Island Pond, and Blue Bank. Many of the old pits can still be found today, hidden by overgrowth or tucked among the hardwoods as anonymous little forest ponds.

Hatcher's field diaries reveal that he collected most of his Arundel dinosaur fossils under extreme weather conditions in the dead of the 1887–1888 winter. He mentions not only the Swampoodle site in the Muirkirk-Contee area of Prince Georges County, but also several locations in nearby Maryland counties and the city of Baltimore. The entire corridor between Baltimore and Washington came to be known as "Dinosaur Alley" for its abundant fossil discoveries.

Hatcher shipped hundreds of bones and teeth back to Marsh at the Peabody Museum, commenting on their scrappy condition: "No two bones or fragments of all that material collected from the Potomac beds in Maryland were found in such relation to one another as to demonstrate that they belonged to the same individual. In any discussion as to the affinities of these various genera and species of small Sauropod dinosaurs, not only the immature nature of the remains upon which they have been based, but also the

FIG. 4.23 Arthur Barneveld Bibbins explored the dinosaur-rich Arundel Clay in Maryland in 1894–1896. (Goucher College Archives)

scattered and disarticulated state in which found, must be constantly borne in mind." Even so, Marsh was able to identify five new dinosaur species from the Hatcher material in 1888 (chapter 8). These were the sauropods *Pleurocoelus nanus* and *Pleurocoelus altus,* the ankylosaur *Priconodon crassus,* and the theropods *Allosaurus medius* and *Coelurus gracilis.* Marsh pronounced the age of the fossils as Late Jurassic, revised by later researchers to Early Cretaceous.

The Hatcher fossils now reside at the Smithsonian's National Museum of Natural History in Washington. They share that home with another great assemblage of Arundel dinosaur material collected by a young Maryland geologist named Arthur Barneveld Bibbins. While a graduate student at the Johns Hopkins University, Bibbins (fig. 4.23) was hired in 1894 as museum curator and instructor in geology and biology at the Woman's College of Baltimore, now known as Goucher College. Bibbins later started his own mineral resources business and did considerable fieldwork for the Maryland Geological Survey and the United States Geological Survey.

Bibbins began collecting from the Arundel Clay in 1894. His first efforts yielded fossil cycad trunks, which had initially been discovered by Philip Tyson in 1859 during the same scouting activity that produced the *Astrodon johnstoni* teeth. Bibbins was adept at finding the cycad fossils, whether in the ground or gathering dust in a local resident's home, and he soon applied his skills to dinosaur bones: "During the summer of 1894 [my] attention was directed to additional saurian fossils which had been found in the same mine explored by Marsh, and at various other points apparently yet unexplored. Several of these new localities were examined with some care in the hope both of adding to our knowledge of the species already known and of the discovery of additional forms. Though the work was attended by much diffi-culty, a considerable collection was obtained in a comparatively short time."

When the Goucher College collection was permanently lent to the U.S. National Museum at the Smithsonian Institution in 1916, the accession records listed "84 identifiable and 59 fragmentary specimens of fossil verte-brates." Among Bibbins's prized dinosaur discoveries was "a tibia, probably of *Allosaurus,* which measures ten inches in width, and thirty-two inches in length, exclusive of the ends, which are lacking." In addition to *Allosaurus,* Bibbins mentions finding specimens of *Pleurocoelus, Priconodon,* and *Astrodon.*

End of an Era

The story of Early Cretaceous dinosaurs on the East Coast was largely writ-ten by 1896, when Bibbins had collected most of his fossils. Only a handful of documented discoveries occurred in the Arundel Clay during the next sev-en decades. Among them were two finds made in Washington, D.C., the only dinosaur fossils ever recorded from the nation's capital.

In 1898 workmen installing a sewer line at First and F Streets SE came upon a vertebra and bone fragments of a large carnivorous dinosaur. The oth-er Washington discovery was made in 1942, when a large *Astrodon* thighbone was uncovered during construction on the McMillan Water Filtration Plant at First and Channing Streets NW. The bone was described by Charles W. Gilmore, curator of vertebrate paleontology at the U.S. National Museum (chapter 9), as belonging to a sauropod "about ten feet high at the hips and 50 or 60 feet long, weighing approximately 10 tons. Apparently, from the geological evidence, it was trapped in a small pond or mudhole and so perished."

These coincidental finds of fossils were typical of dinosaur paleontology on the East Coast in the first half of the twentieth century. Scientific attention had shifted to the highly productive fossil fields of the western United States and Canada, and collecting languished in the eastern states. The quarrying industries that had opened so many prime exposures in dinosaur fossil-bear-ing sediments in the 1800s gradually shut down.

It wasn't until the 1960s that scientific collecting was revived on the East Coast, with a subsequent burst of activity in the Cretaceous sites of Maryland, New Jersey, Delaware, and the Carolinas and new interest in the Late Trias-sic/Early Jurassic footprints of the Connecticut Valley, Pennsylvania, New Jersey, and Virginia (chapter 9).

5

DINOSAURS
OF THE LATE TRIASSIC

The reign of Dinosauria began in the Late Triassic Period, some 230 Ma. The time must have been right for the arrival of these new life forms, because dinosaurs soon had a nearly global distribution, including the East Coast of North America. They also had begun the great diversification that produced all the major dinosaur groups (chapter 1). In the Late Triassic there were already theropods, prosauropods, and the earliest ornithischians.

Although fossil evidence for these early dinosaurs is most abundant in South America and Africa, we find their remains in Upper Triassic rocks around the world. All major landmasses were then joined in the supercontinent Pangaea (chapters 2 and 3), and the barriers to terrestrial migration that come from expanding oceans and rising mountain ranges were fewer and farther between. Unlike their descendants in later periods of the Mesozoic, the dinosaurs of the Late Triassic could move freely throughout Pangaea.

Late Triassic: The Pangaea Story

Before entering the East Coast world of the Late Triassic, let's take stock of the earliest dinosaurs, their diversity, and their geographic distribution (called paleobiogeography). The best record comes from Gondwana, the southern half of Pangaea, which roughly corresponds to today's Southern Hemisphere continents. The Gondwana dinosaur remains of the Late Triassic consist of body fossils—skulls, skeletons, teeth—and footprints and trackways.

It appears that dinosaurs originated in what is now South America. Their closest relatives (*Lagosuchus:* chapter 10) are found there, as are the most primitive members of Dinosauria itself. These latter animals include an early theropod called *Herrerasaurus* ("Herrera reptile," named for Don Victorino Herrera, who led paleontologists to its discovery in Argentina); one of the earliest ornithischians, *Pisanosaurus* ("Pisano reptile," named for Argentine paleontologist Juan A. Pisano); and a few prosauropods from Argentina: *Riojasaurus* ("Rioja reptile," named for La Rioja province) and *Coloradisaurus* ("Colorados reptile," named for the Los Colorados Formation). Elsewhere in Gondwana—in Africa, Australia, and India (which was then attached to Antarctica)—the dinosaur fossil record consists largely of prosauropod and theropod remains, both skeletons and footprints.

Late Triassic dinosaurs were also widely distributed across Laurasia, the northern half of Pangaea that included present-day North America, Europe, and Asia. In China there were diverse prosauropods and a few theropods. Farther to the west, Europe (Great Britain, Belgium, France, Germany, Switzerland, Italy, and Poland) was also populated by early dinosaurs. Among the better-known prosauropods in Europe were the German *Plateosaurus* ("flat reptile") and *Thecodontosaurus* ("socket-toothed reptile") from England. Primitive theropods included *Liliensternus,* named in honor of German paleontologist H. R. von Lilienstern.

Closer to the East Coast, one of the best records of Late Triassic dinosaurs in North America comes from the Four Corners region of the United States (Arizona, New Mexico, Colorado, Utah), as well as adjacent parts of western Texas. At Ghost Ranch, New Mexico, an abundance of skeletal remains of the ceratosaurian theropod known as *Coelophysis* ("hollow form") was found in the Chinle Formation. Several theropod footprint localities have been discovered in Utah, Colorado, and Texas. But prosauropods are rare in this part of the world: perhaps this is one region of Pangaea that escaped their great wanderings.

East Coast Dinosaur Distribution

The Late Triassic (and Early Jurassic) are better documented than any other Mesozoic periods on the East Coast of North America (table 5.1). The

TABLE 5.1   Late Triassic Dinosaurs of the East Coast

Dinosaur	Locality	Geologic Formation	Description[a]
Ornithischia			
Galtonia gibbidens	Pennsylvania	New Oxford Formation	S, small primitive plant eater, formerly *Thecodontosaurus gibbidens*.
Pekinosaurus olseni	North Carolina	Pekin Formation	S, small primitive plant eater.
Atreipus	Pennsylvania, Virginia, New Jersey, and Nova Scotia	Lockatong, Gettysburg, Cow Branch, Passaic, and Wolfville formations	F, three-toed hind prints, four-toed fore prints. Quadrupedal trackmaker about 1–2 m in length, less than 1 m high at the hips.
Gregaripus	Virginia	Balls Bluff Siltstone	F, small three-toed hind prints. Trackmaker 0.5 m at hips, 1.5 m long.
Saurischia			
Grallator	Pennsylvania, Virginia, New Jersey, New York, and Nova Scotia	Lockatong, Gettysburg, Cow Branch, Passaic, Stockton, and Blomidon formations	F, small three-toed hind prints. Bipedal trackmaker 0.5 m at hips, 2–3 m long.
Kayentapus	Virginia	Balls Bluff Siltstone	F, medium sized three-toed hind prints. Bipedal trackmaker 1 m at hips, 3–4 m long.
Agrestipus	Virginia	Balls Bluff Siltstone	F, small four-toed hind prints. May be a prosauropod.

[a]S = skeletal remains; F = footprints.

Newark Supergroup sediments that stretch from South Carolina to Nova Scotia (chapter 3) provide a wealth of fossil evidence of land animals, as well as lake-dwelling fishes, amphibians, and reptiles. The oldest Newark Supergroup rocks appear to be in central Virginia, but these sediments contain no record of dinosaurs, and paleontologists are fairly certain none lived here at that time.

The first glimmer of dinosaurs on the eastern seaboard appears about

225 Ma. Most of this evidence comes in the form of abundant footprints and trackways, but there are also fragmentary skeletal remains. From south to north, the major Upper Triassic sediments that contain dinosaur fossils are as described below.

North Carolina (Pekin Formation)

The Pekin Formation of the Deep River Basin in North Carolina has yielded evidence of a primitive ornithischian dinosaur known as *Pekinosaurus olseni*. Nondinosaur fossils from the Pekin include the crocodile-like phytosaur *Rutiodon* and several armored, plant-eating aetosaurs. Also from the Pekin are the archosaur footprints *Brachychirotherium* and *Apatopus*.

Virginia/North Carolina (Cow Branch Formation)

The record of early dinosaurs improves in the adjacent western part of Virginia, in the Cow Branch Formation of the Dan River Basin. Here, at Leaksville Junction on the border of Virginia and North Carolina, we find what may be the oldest dinosaur footprints on the East Coast (chapter 9). This early record of isolated prints and trackways provides evidence for both theropods and ornithischians. The three-toed hind-foot prints known as *Grallator* were made by a bipedal, meat-eating theropod. The quadrupedal *Atreipus* footprint also appears, as three-toed hind-foot tracks and much smaller three- or four-toed forefoot prints. These abundant *Atreipus* tracks are thought to have been made by a primitive ornithischian dinosaur. Nondinosaur fossils from this Upper Triassic site include skeletal remains of the phytosaur *Rutiodon* and well-preserved specimens of *Tanytrachelos*, a small, water-dwelling, lizardlike reptile.

Virginia (Manassas Sandstone)

Sediments of the Culpeper Basin near Manassas in Fairfax County, Virginia, have produced reptile skeletal remains and footprints from at least three sites. Dinosaur footprints include *Grallator*. The archosaur tracks *Apatopus*, *Brachychirotherium*, and *Chirotherium* have also been uncovered. Nondinosaur skeletal material includes phytosaur and fish remains.

Virginia (Balls Bluff Siltstone)

The Balls Bluff Siltstone of the Culpeper Basin has yielded some of the world's finest and most abundant tracks and trackways of Late Triassic dinosaurs. Two print-laden bedding planes at a stone and gravel quarry near

Culpeper (chapter 9) have produced the theropod footprints *Grallator* and *Kayentapus*, the prosauropod track *Agrestipus*, and the small ornithischian print *Gregaripus*. Nondinosaur footprints at the Culpeper site include aetosaur tracks. Another site in the Balls Bluff Siltstone, near Manassas National Battlefield, has yielded a single *Grallator* track.

Pennsylvania (New Oxford Formation)

The only skeletal remains of dinosaurs found in the Gettysburg Basin—and in all of Pennsylvania—have come from the New Oxford Formation. *Galtonia gibbidens*, an ornithischian dinosaur, is represented by several teeth described in 1878 from Emigsville, Pennsylvania. Nondinosaur material from the New Oxford Formation includes skeletal remains of the phytosaur *Rutiodon*, found along Little Conewago Creek at Zions View in York County. Two metoposaur (large amphibian) specimens were collected from a copper mine near Emigsville.

Pennsylvania/Maryland (Gettysburg Formation)

This formation of the Gettysburg Basin has yielded *Atreipus* footprints, including finds in 1937 at York Springs, Pennsylvania, and in 1889 at Goldsboro, Pennsylvania. In 1895 small *Grallator* tracks were discovered in Gettysburg deposits at Emmitsburg, Maryland (chapter 4). Nondinosaur footprints include *Brachychirotherium*, *Rhynchosauroides*, and *Apatopus*.

Pennsylvania/New Jersey/New York (Stockton Formation)

Dinosaur fossils are rare in the oldest layer of the Newark Basin. This fluvial, or floodplain, deposit of mudstones, shales, and sandstones accumulated some 230 to 225 Ma. *Grallator* tracks have been found in upper Stockton beds at Nyack Beach State Park in Haverstraw, New York, along with other reptile prints. Fossil remains from the Stockton Formation include metoposaurs, coelacanth fishes, and phytosaurs. The oldest vertebrate fossil yet found in either the Newark or Hartford Basin is the lower jaw of a primitive amphibian known as *Calamops*, from Holicong, Pennsylvania, in the early 1900s.

Pennsylvania/New Jersey (Lockatong Formation)

This formation of gray siltstones was the bottom of huge lakes that covered much of central New Jersey and parts of Pennsylvania during the Late Triassic. *Grallator* tracks appear in Lockatong sediments, as do *Atreipus* footprints from Gwynedd, Arcola, and Graterford, Pennsylvania. The Lockatong has also produced abundant nondinosaur fossils, including metoposaurs and coelacanths (*Diplurus*) and other bony fishes. Phytosaurs such as *Rutiodon* are well represented by skeletal remains, including a small skull found at the Granton Quarry in North Bergen, New Jersey. The phytosaur footprints named *Apatopus* are present in the Lockatong, as well as the primitive archosaurian reptile tracks *Rhynchosauroides*, *Brachychirotherium*, and *Chirotherium*. The remains of a gliding lizard from North Bergen were described in 1966 by Edwin Colbert as *Icarosaurus siefkeri*.

Pennsylvania/New Jersey (Passaic Formation)

The lower part of the Passaic Formation (formerly known as the Brunswick Formation) of the Newark Basin preserves fossil evidence of the great extinction event that ended the Triassic. Dinosaurs ranged the Newark Basin throughout Passaic time, leaving many tracks. One of the historically most productive Passaic footprint sites was the Smith Clark Quarry in Milford, Hunterdon County, New Jersey, which yielded large and small *Grallator* tracks and the primitive crocodile-like print known as *Otozoum*. More *Grallator* tracks were found in New Jersey at Newark, Hackensack Meadows, and Lyndhurst-Rutherford, and in Pennsylvania at Schwenksville. *Atreipus* tracks have been found in New Jersey at Milford and Frenchtown, and in Pennsylvania at Graterford.

Fossils of nondinosaurian reptiles from the Passaic Formation provide the best evidence for the Triassic/Jurassic extinction. Many species and higher groups of plants and animals vanished or were reduced in number at the beginning of the Jurassic (chapter 6). The vertebrate tracks or remains found in the lower part of the Passaic—but not in the overlying rock units—include metoposaurs, the phytosaur *Rutiodon*, the archosaur *Stegomus*, the primitive archosaur tracks *Apatopus*, *Chirotherium*, *Brachychirotherium*, and *Rhynchosauroides*, and the procolophonid reptile *Hypsognathus*.

Nova Scotia (Wolfville Formation)

On the cliffs and beach southeast of Paddy's Island in Kings County, Nova Scotia, collectors have discovered *Atreipus* footprints in the Wolfville Formation of the Fundy Basin (chapter 9). Dinosaur skeletal remains include the tooth of an indeterminate primitive ornithischian and the bones of a prosauropod. Nondinosaur footprints include *Rhynchosauroides*, plus several *Brachychirotherium* trackways. Just below the horizon at Paddy's Island, the

Wolfville has also yielded the skull and partial skeleton of an unidentified primitive reptile, possibly *Hypsognathus*. Other skeletal remains provide evidence for metoposaurs and a small aetosaur.

Nova Scotia (Blomidon Formation)

At Red Head, on the north shore of St. Mary's Bay near Rossway, Digby County, Nova Scotia, the Blomidon Formation of the Fundy Basin has produced large and small *Grallator* footprints. Nondinosaur fossils include *Rhynchosauroides* footprints and a snout fragment of the phytosaur *Rutiodon*.

East Coast Dinosaurs of the Late Triassic

These are the occurrences of Late Triassic dinosaurs throughout the East Coast states and the Maritime province of Nova Scotia. Thanks to the efforts of such pioneers as Edward B. Hitchcock, Richard Swann Lull, and Edward D. Cope—and the modern-day research of Donald Baird, Edwin H. Colbert, Paul E. Olsen, Hans-Dieter Sues, Neil H. Shubin, Robert E. Weems, Nicholas C. Fraser, Peter M. Galton, Adrian P. Hunt, Spencer G. Lucas, Ronald J. Litwin, Robert M. Sullivan, Shaymaria M. Silvestri, and others—we can paint a fairly complete picture of the Late Triassic dinosaurs of the East Coast. Let's begin with the ornithischians.

Ornithischia: Bones, Tracks, and Behavior

The scene is the Late Triassic on the East Coast. Hidden in the undergrowth of ferns and cycads, a small, nimble dinosaur browses on succulent leaves and an occasional fruit. At times it rears up on its hind legs to look around or sniff the air for predatory dinosaurs. When the coast is clear, the plant eater dashes out of its forest cover and sprints across the adjacent mudflat to sample leafy morsels in another grove of trees.

FIG. 5.1 A front tooth of the primitive ornithischian *Galtonia gibbidens,* from the Late Triassic of Pennsylvania (after Hunt and Lucas 1994).

This image of the earliest ornithischian dinosaurs of the eastern seaboard comes to us imperfectly through the fossils of *Galtonia gibbidens* of south-central Pennsylvania, *Pekinosaurus olseni* from central North Carolina, and the makers of *Gregaripus* and the ubiquitous *Atreipus* tracks.

Galtonia gibbidens is represented by several teeth collected in the mid-1800s from near Emigsville, Pennsylvania. The teeth (fig. 5.1) were originally described in 1878 by Edward D. Cope as the prosauropod dinosaur *Thecodontosaurus gibbidens* ("socket-toothed reptile, protuberant tooth"), based on similarities with *Thecodontosaurus antiquus* ("old socket-toothed reptile") from England. In 1994 Adrian P. Hunt and Spencer G. Lucas named the teeth *Galtonia gibbidens* ("Galton's protuberant tooth") for Peter M. Galton of the University of Bridgeport in Connecticut, a specialist on early dinosaurs who first recognized their ornithischian affinity (chapter 9).

Pekinosaurus olseni ("Olsen's Pekin reptile") was also named by Hunt and Lucas in 1994 based on several ornithischian teeth collected near the town of Pekin, North Carolina. The species name honors Paul E. Olsen of the Columbia University Lamont Doherty Earth Observatory (chapter 9).

The teeth of *Galtonia* and *Pekinosaurus* tell us that we are dealing with small, primitive ornithischian dinosaurs. These teeth are quite small, no more than 5 to 6 millimeters (0.18 to 0.22 inch) high and broadly triangular. Such features, found in other primitive ornithischians as well, characterize the earliest members of this diverse group of plant-eating dinosaurs.

Ornithischian footprints and trackways are abundant on the East Coast, in contrast to the rare skeletal remains. Particularly common is the print known as *Atreipus* ("Atreus's foot," named for Atreus Wanner, a high-school principal and discoverer of many Late Triassic tracks in York County, Pennsylvania, in the late 1800s). There are several kinds of the quadrupedal *Atreipus*, among them *Atreipus milfordensis* and *Atreipus acadianus* (fig. 5.2). Among the oldest dinosaur footprints on the East Coast are *Atreipus* tracks discovered at Leaksville Junction on the border of Virginia and North Carolina, currently under study by Olsen and Nicholas C. Fraser, curator at the Virginia Natural History Museum in Martinsville. *Atreipus* is so common and sufficiently restricted in time—known only from about a 10 million year period—that it is often used as an index fossil for part of the Late Triassic.

A less prolific ornithischian trackmaker is *Gregaripus bairdi* ("Baird's sociable foot," named for Donald Baird by Robert E. Weems of the U.S. Geological Survey in 1987). These tracks have been found in only one location: the upper of two footprint horizons at the stone and gravel quarry near Culpeper,

Virginia. *Gregaripus* tracks are blunt, three-toed, and small, less than 10 centimeters (4 inches) long.

How do we know *Atreipus* and *Gregaripus* were ornithischians? The three-toed hind-foot prints of *Atreipus* have a long central toe (called digit III) and two side toes (digits II and IV) that are "subequal": shorter than the central toe but equal in length to each other. By noting the pattern of pads along each toe, we know there are three toe bones (phalanges) in digit II, four in digit III, and five in digit IV. This gives a digital formula of ?-3-4-5-? (digits I and V, if present, were not preserved as prints).

But this footprint pattern is very similar to that of the most abundant dinosaur footprint on the East Coast: the bipedal theropod tracks known as *Grallator* (refers to Grallae, the group of birds that includes herons and storks). To distinguish ornithischian and theropod tracks, it is important to look beyond the hind-foot prints. And in the case of *Atreipus* we find impressions of the front feet—small, clawed, four-toed—which, according to Olsen and Baird, tell us we're dealing with a quadrupedal ornithischian rather than a bipedal theropod.

From the size and shape of the front and hind tracks, we can assess the size and shape of the animal that made *Atreipus* footprints. First, the trackmaker was less than 1 meter (3.3 feet) high at the hips and, by comparison with other ornithischian dinosaurs known from skeletal material, 1 to 2 meters (3.3 to 6.6 feet) long. Because of the abundance of forefoot prints, *Atreipus* must

FIG. 5.2 Front- and hind-foot prints of *Atreipus,* thought to have been made by a primitive ornithischian, from the Late Triassic of the East Coast (after Olsen and Baird 1986).

have ambled around on all fours, although the size of these prints suggests that the front legs were smaller than the hind legs. The pattern of pads on the forefoot prints also tells us that the digital formula of the hand is 2-3-4-3-? (the fifth digit was not pressed into the sediment). This pattern, as well as the relative sizes of the toe bones, is right on the money for an early quadrupedal ornithischian dinosaur.

Although few specimens of *Atreipus* consist of more than a few prints arranged in a trackway, those of *Gregaripus* are known from more than a dozen trackways at Culpeper, including one that is thirty-four prints long. Consisting solely of hind-foot prints, the *Gregaripus* tracks were likely made by an early ornithischian that stood about 0.5 meter (1.6 feet) high at the hips and was 1.5 meters (5 feet) from snout tip to tail end. From the trackways, it is estimated that the *Gregaripus* trackmaker was moving at up to 3 meters a second (6.7 miles an hour), making this small ornithischian a fairly rapid runner.

We also learn from the Culpeper prints that the *Gregaripus* trackmaker was probably a sociable beast. The tracks reveal dinosaurs apparently traveling in the same direction in a structured fashion: smaller individuals surrounded by larger dinosaurs, almost as if youngsters were being protected by adults as they ran across the mudflat. If this is true, then it is probably also true that these primitive ornithischians lived in large, well-organized family groups.

To learn more about the Late Triassic ornithischians of the East Coast, we must look to better-known primitive ornithischians from other parts of the world. One example is the Early Jurassic *Lesothosaurus* ("Lesotho reptile"), named for the small country in southern Africa where it was discovered (fig. 5.3).

This dinosaur has assumed considerable importance in our understanding of the evolution of ornithischians. Research by Paul C. Sereno of the University of Chicago and others has shown that *Lesothosaurus* represents one arm of the initial split of this great group of plant-eating dinosaurs. The other arm is the group known as Genasauria ("cheeked reptiles"), which includes all remaining ornithischians: stegosaurs, ankylosaurs, pachycephalosaurs, ceratopsians, and ornithopods. All—but not *Lesothosaurus*—had acquired important anatomical "inventions" such as muscular cheeks and a spout-shaped front to the lower jaws.

Measuring 1 meter (3.3 feet) long, *Lesothosaurus* was identified and christened in 1978 by Galton. He recognized that *Lesothosaurus* had been incorrectly grouped with ornithopods based on its primitive nature, and it is now widely considered the most primitive of all known ornithischians. Because

FIG. 5.3 *Lesothosaurus*, a primitive ornithischian from the Early Jurassic of southern Africa.

its skeletal remains are very well preserved, *Lesothosaurus* is an excellent candidate for comparison with the rare and poorly known ornithischian dinosaurs from the Late Triassic of the East Coast.

Four features characterize ornithischian dinosaurs. First, the lower jaws were capped in front by a horn-covered bone called the *predentary* that improved their ability to nip leaves and fruits from shrubs and low-hanging tree branches. Second, along the back and out onto the tail was a trellislike arrangement of long, narrow bony tendons that counterbalanced the front of the body and the tail over the hips. Third, the pubic bone was rotated backward to lie adjacent to the ischium (the hipbone), permitting a larger gut for more efficient fermentation of plant food. Fourth, a bony rod called a *palpebral bone* crossed the upper part of the eye cavity, probably encased in the upper eyelid.

Ornithischians had slender, long hind limbs and short forelimbs with small hands, and the tips of the five stubby fingers ended in small claws. The tail was long and pointed straight out from the rear for better balance when the animal reared up on its hind legs. Nevertheless, ornithischians probably rested and walked slowly on all fours, picking up the front limbs only when running fast. The neck was long and flexible, and the head was rather small and triangular in profile, with a large eye socket and a large chamber for the muscles that worked the jaws. In front of the eye socket was an opening for an air chamber much like our own sinuses.

The small teeth of primitive ornithischians are often leaf shaped. From this shape and the wear that is sometimes found on the tooth crown, we can conclude that they were fairly good chewers. Ornithischians ate the foliage of ferns, seed ferns, conifers, and cycads. Competing for this food were other herbivores, especially the aetosaurs. These were 3 to 4 meters (10 to 13 feet) long, heavily armored, pig-snouted animals like *Stegomus* ("bent covering") that walked close to the ground on all fours and may have fed on roots and tubers.

Ornithischians—with their small, lightly built skeletons and agility—probably relied on speed and coordination to avoid being caught by large predatory dinosaurs. In the Late Triassic along the East Coast, that means our next subject: the makers of the abundant *Grallator* footprints.

Theropoda: Tracking and Attacking in the Triassic

It is unfortunate that no skeletal fossils belonging to theropod dinosaurs have been discovered in Upper Triassic sediments along the East Coast. Yet their abundant footprints have provided important clues to theropod diversity and paleobiology in this period.

The tracks—all from the hind foot, which indicates bipedality—have been given a wide variety of names, but they all bear many of the same diagnostic

FIG. 5.4 Hind-foot track of *Grallator,* thought to have been made by a ceratosaurian theropod, from the Late Triassic and Early Jurassic of the East Coast (after Lull 1953).

features: three toes whose pads reveal the number of toe bones, terminal claw marks; and the impression of a heel and the trace of a fourth, rearward-pointing toe. The middle, third digit is always the longest, while the side toes (digits II and IV) are subequal. With all these features in common, it is often the size of the prints that determines the names applied to them.

We have already introduced *Grallator,* a small theropod footprint 5 to 15 centimeters (2 to 6 inches) long, first documented by Edward Hitchcock in 1858 (fig. 5.4). The three-toed *Grallator* prints are abundant and widely distributed. Several kinds of *Grallator* tracks have been named, including *G. cursorius* and *G. tenuis.* Both small and large *Grallator* prints appear at many East Coast localities.

Often found with *Grallator* are prints named *Anchisauripus* ("Anchisaurus foot"), first described by Richard Swann Lull in 1904 as the probable track of the prosauropod dinosaur *Anchisaurus* ("near reptile"). Specialists now believe that *Anchisauripus* tracks were made not by a prosauropod but by a theropod, as were those of *Grallator.* Indeed, it is probable that *Anchisauripus* tracks—15 to 25 centimeters (6 to 10 inches) long—are the footprints of large individuals of the *Grallator* trackmaker. The digital formula of the tracks of *Grallator,* large or small, is ?-3-4-5-?.

The third of these Late Triassic footprints is *Kayentapus,* found on the lower level of the Culpeper quarry. *Kayentapus* ("Kayenta foot," named for the Kayenta Formation of Arizona, where similar tracks of Early Jurassic age were found) seems to come from a different theropod trackmaker. The East Coast *Kayentapus* tracks are large—30 centimeters (11 inches) long—with a much longer, claw-tipped middle toe and subequal, diverging side toes. The digital formula of *Kayentapus* is the same as for *Grallator.*

Judging from the size of the prints, it appears that *Grallator* was about 2 to 6 meters (7 to 20 feet) long, whereas the *Kayentapus* trackmaker was probably 3 to 4 meters (10 to 13 feet). Based on their form and size, the maker of *Kayentapus* tracks has been likened to *Dilophosaurus* ("two-crested reptile"), a 6 meter (20 foot) long theropod known from the Early Jurassic of Arizona.

In contrast, the maker of *Grallator* prints is thought to have been something like *Coelophysis* (fig. 5.5), which is older (Late Triassic) and smaller (3 meters, or 10 feet) than *Dilophosaurus.* Another possible candidate for the *Grallator* trackmaker is the Early Jurassic *Syntarsus* from Arizona and southern Africa, similar in size and form of the hind foot.

If we are right in thinking that these footprints were made by theropods like *Coelophysis, Syntarsus,* or *Dilophosaurus,* then we need to know more about

FIG. 5.5 *Coelophysis,* a ceratosaurian theropod from the Late Triassic of the southwestern United States.

these dinosaurs. They are members of a group of primitive theropods called ceratosaurs ("horned reptiles"). The recognition of this group is a by-product of the drive to understand the origin of birds (chapter 10). Ceratosauria— named after the Late Jurassic *Ceratosaurus* from the Western Interior of the United States—was first recognized in 1986 by Jacques A. Gauthier of the California Academy of Sciences and subsequently studied in detail by Gauthier and Timothy Rowe of the University of Texas at Austin.

As a group, ceratosaurs have a number of unique features: neck vertebrae with two pairs of hollow spaces called *pleurocoels;* an ilium (upper hipbone) that fuses in adulthood with the rest of the hipbones (pubis and ischium); and a fibula—the bone that runs along the side of the tibia, or shinbone—that flares at its lower end.

Within Ceratosauria, *Dilophosaurus* is closely related to *Coelophysis, Syntarsus,* and *Liliensternus* from Germany. All these animals were small by later theropod standards, approximately 2 to 3 meters (7 to 10 feet) from tip of nose to tip of tail. Most ceratosaurs were quite slender, though the hind legs were powerful. There is no doubt that these bipedal dinosaurs were swift and agile.

Quite long armed, ceratosaurs had hands equipped with three strong claw-tipped fingers that could function in grasping. Although these clawed hands would have made formidable weapons, it was the ceratosaurs' sharp, recurved teeth that were most dangerous to prey. Both *Dilophosaurus* and *Syntarsus* had a strikingly visual crest on top of the head made from two thin ridges of bone. This cranial crest may have functioned in behavioral displays, territorial disputes between males, or mating rituals.

Turning back to the East Coast, what more do *Grallator* and *Kayentapus* tell us about the Late Triassic ceratosaurs of the region? To answer this question, we look to the trackways on both lower and upper levels of the Culpeper Quarry. The most common prints along the lower level, arranged in as many as twenty trackways, are of *Kayentapus*, while the upper level contains nearly a half dozen *Grallator* trackways.

The Culpeper trackways reveal the swift pace of these theropods, thanks to the research efforts of Robert E. Weems (chapter 9) of the U.S. Geological Survey in Reston, Virginia. For example, the maker of *Grallator* tracks was capable of traveling at 3.5 to 4.4 meters a second (7.8 to 9.8 miles an hour), while the *Kayentapus* trackmaker apparently moved at a blazing 8.4 meters a second (about 19 miles an hour). Clearly speed—particularly speed in the attack—was an important aspect of the biology of these animals.

We can also identify the probable prey of the *Grallator* and *Kayentapus* trackmakers. These include the ornithischians *Pekinosaurus, Galtonia,* and the fleet-footed maker of the *Atreipus* tracks, as well as insects, amphibians, and lizards. The theropods also had their own enemies in such ferocious, four-footed archosaurs as the phytosaur *Rutiodon* ("digging tooth") and the makers of *Batrachopus* ("frog foot") and *Brachychirotherium* ("short-gloved beast") tracks.

Prosauropods: The Beginnings of Big

Although they are known elsewhere as the first large browsers of the Mesozoic (chapters 6), the Late Triassic prosauropods are poorly represented on the East Coast. There are only a few badly preserved bony specimens from Nova Scotia and some indistinct hind-foot prints from Culpeper that may have been made by a prosauropod known as *Agrestipus hottoni* ("Hotton's clumsy foot"). This footprint species was named by Weems in 1987 for Nicholas Hotton III, a specialist on late Paleozoic and early Mesozoic tetrapods and former curator of fossil reptiles at the Smithsonian Institution in Washington, D.C.

The ducklike hind-foot prints of *Agrestipus* reveal three, and sometimes four, blunt toes. Originally suggested by Weems to have been made by sauropod dinosaurs, they are probably prosauropod prints. This interpretation is more consistent with the fact that the earliest sauropod skeletal remains from elsewhere in the world are Early Jurassic in age, not Late Triassic.

If *Agrestipus* belongs to a prosauropod, then these footprints are the best record we have of these animals. They confirm the presence, though extremely rare, of prosauropods in this part of the world during the Late Triassic. And they paint a picture of a slow-walking bipedal animal, ambling across the mudflats of Culpeper at 1 meter a second (2.2 miles an hour).

The Great Triassic Extinction

By the end of the Triassic, dinosaur diversity had grown considerably, if a bit lopsidedly. We had both the small, meat-eating theropods and large, plant-eating prosauropods. And yet we had only the barest beginnings of ornithischian dinosaurs. The great flurry of evolution among these plant eaters must await the Middle Jurassic, some 30 to 40 million years in the future.

As befits the huge Pangaean landmass the dinosaurs inherited at the end of the Triassic, there was a great deal of interaction between these and other animals on a nearly worldwide scale. Dinosaurs, especially theropods and prosauropods, were distributed from Argentina to China and from Australia to England.

But much of this was to end at the close of the Triassic, a time when many of these dinosaurs and a host of other land-living creatures disappeared in rapid succession. The victims included the small, plant-eating procolophonids (early predecessors to modern reptiles), the crocodile-like phytosaurs, the squat rhynchosaurs (piglike distant relatives of dinosaurs and crocodilians), and great numbers of dicynodont and cynodont therapsids (herbivorous and carnivorous mammal-like reptiles), as well as several major groups of predatory archosaurs. In the plant realm, there was a decimation of seed fern floras, including horsetails, ferns, cycads, ginkgoes, and conifers.

The causes of the Late Triassic extinctions—there may have been two closely spaced events—are a matter of some controversy. Recent opinion is that they probably had very little to do with the rise of superior animals and plants in competition with less successful forms. Instead, Michael K. Benton of Bristol University in England has suggested that the extinctions may have been linked to rapid climate changes.

These changes would have had major effects on terrestrial plants and the animals that depended on them. The extinctions may have progressed as a series of interrelated crises. First, the seed fern floras succumbed to climatic conditions, then the herbivores that fed on them became extinct, then the predators that preyed on the herbivores died off. Far from being a long-term competitive takeover, the rapid loss of the dominant land-living vertebrates set the stage for opportunistic evolution at the end of the Triassic.

A more dramatic, and more controversial, explanation for the Late Triassic extinctions involves asteroid impact. These extraterrestrial objects may be the single most important cause of profound and rapid death ever inflicted on the life of Earth. As in the mass extinction at the end of the Cretaceous (chapter 10) that killed off the dinosaurs, it may also be possible to connect the Triassic events with a specific impact crater.

Paul Olsen and his colleagues initially identified the Triassic "smoking gun" as Manicouagan Crater in northern Quebec, Canada. Manicouagan is 70 kilometers (43 miles) in diameter, nearly as large as the suspected impact site for the Late Cretaceous extinction: the Chicxulub Crater on the Yucatán Peninsula in southeastern Mexico. Unfortunately the most recent dating of the Manicouagan cratering is too old for the Triassic-Jurassic boundary. But Olsen and his colleagues have not given up. Other signatures of an asteroid impact at the boundary are now being sought: concentration of iridium, microtektites, and shocked quartz, all of which attest to an intensely violent collision of an asteroid with Earth (chapter 10).

If it's true that something from "out there" devastated life on Earth at the end of the Triassic, then perhaps this turnover in the make-up of terrestrial faunas—which included the earliest saurischian and ornithischian dinosaurs—sets the stage for evolutionary diversification throughout the rest of the Mesozoic.

6

DINOSAURS
OF THE EARLY JURASSIC

The descendants of the earliest dinosaurs were among the lucky few to survive the great extinction at the end of the Triassic Period. Many groups of amphibians, procolophonids, and primitive archosaurs didn't make it, but several other groups—including pterosaurs, turtles, and dinosaurs—somehow managed to cross over into the Jurassic.

What sort of world greeted these survivors? Northern Africa and Europe to the east and North America to the west continued to slowly pull apart to form the ever-expanding North Atlantic Ocean (chapter 3). The rift valleys that had opened on the continental margins contained a great chain of lakes, and it was around these lakes that East Coast dinosaurs lived.

But there were some new members of the dinosaur menagerie inhabiting these landscapes, as well as carryovers like the *Grallator* trackmaker. To understand the Early Jurassic dinosaurs of the eastern seaboard, we need a global picture of the dinosaurs that lived during this interval of the Mesozoic.

Early Jurassic: Breaking up Is Hard to Do

Things were changing in the world of dinosaurs throughout the Early Jurassic. The groups that had their origin in the Late Triassic—ceratosaurian theropods, prosauropods, and primitive ornithischians—show a great deal of evolutionary success.

In Gondwana, the southern part of Pangaea, Africa provides a bonanza of Early Jurassic dinosaur skeletal remains and footprints. Here are primitive sauropods such as *Vulcanodon* ("volcano tooth"), the ceratosaurian theropod *Syntarsus*, and the prosauropod *Massospondylus*. In addition to these saurischians, southern Africa sports a host of ornithischian dinosaurs, including the primitive *Lesothosaurus* and the earliest of ornithopods, a group called heterodontosaurids.

Elsewhere in the Southern Hemisphere, only Antarctica and India have provided a record of Early Jurassic dinosaurs. In late 1990, in the windswept mountains of frigid and remote central Antarctica, William R. Hammer and William J. Hickerson of Augustana College in Illinois discovered the remains of a large prosauropod and a new theropod, the latter now called *Cryolophosaurus* ("cold-crested reptile"). India yielded its Early Jurassic dinosaur remains more than twenty years ago, including *Barapasaurus* ("large-legged reptile"). This primitive sauropod provides our best evidence for the emergence of these largest of all terrestrial plant eaters (chapter 1).

The Early Jurassic dinosaurs of Laurasia, the northern part of Pangaea, can be found in a great swath from China to the United States. In Asia the best record comes from southern and central China. This region has yielded magnificent skeletons of a number of prosauropods, a primitive sauropod, and a small theropod. Theropod footprints have been discovered in this same region of China; Afghanistan and Iran have yielded a sparse record of similar tracks, as well as prints that may have been made by early sauropods and ornithopods.

From Europe come some intriguing and often well-preserved Early Jurassic dinosaurs. These include the primitive ornithischian *Emausaurus* ("Emau reptile," named for Ernst-Moritz-Arndt-Universität in Greifswald, Germany); the primitive sauropod *Ohmdenosaurus* ("Ohmden reptile," named for the town of Ohmden, Germany); a ceratosaurian theropod named *Sarcosaurus* ("flesh reptile"); and *Scelidosaurus* ("limb reptile"), an early relative of both stegosaurs and ankylosaurs.

In North America, the desert southwest of the United States was dominated (as it was in the Late Triassic) by ceratosaurian theropods like *Dilophosaurus* and *Syntarsus*. But unlike the Late Triassic, there were also prosauropods like *Ammosaurus* ("sand lizard") and a yet unnamed species of

Massospondylus, as well as the ornithischian *Scutellosaurus* ("small shield rep-tile") and a yet to be described heterodontosaurid ornithopod.

East Coast Dinosaur Distribution

The great chain of rift valleys in the East continued to be active during the Early Jurassic. Terrestrial sediments in which dinosaur fossils have been found are well preserved in central and northern New Jersey, in the Connecticut Valley of Connecticut and Massachusetts, and on the shores of the Bay of Fundy in Nova Scotia. In addition, there are great quantities of fossil fishes, evidence of thriving aquatic communities in the Early Jurassic rift valley lakes. From south to north, the major Lower Jurassic sediments that contain dinosaur fossils are as described below.

Virginia (Midland Formation)

An excellent dinosaur footprint site in the Midland Formation of the Culpeper Basin was discovered about 1920 at Oak Hill, the former estate of President James Monroe in Aldie, Virginia (chapter 4). The Oak Hill dinosaur tracks are *Grallator* and the large theropod prints known as *Eubrontes.* Non-dinosaur prints include the primitive crocodile-like *Batrachopus.*

New Jersey (Passaic Formation)

The upper 100 meters (330 feet) of the Passaic Formation have produced a modest footprint assemblage that includes both small and large *Grallator* prints.

New Jersey (Feltville Formation)

The Feltville Formation is the oldest Newark Basin formation that is entirely Early Jurassic in age. These sandstones cover a wide area of the basin and preserve the first appearance in the Newark Basin of the ornithischian footprint *Anomoepus.* Feltville dinosaur footprints also include *Eubrontes* and large and small *Grallator,* plus the archosaur tracks *Batrachopus.*

New Jersey (Towaco Formation)

The Towaco Formation has produced most of the Jurassic dinosaur footprints in the Newark Basin. A prime exposure is at the Riker Hill Quarry near Roseland, New Jersey (chapter 9). The footprints include *Eubrontes* and large and small *Grallator. Anomoepus* is also present at Riker Hill, as well as many small tracks that may have been made by babies of *Grallator* and *Anomoepus*. Among the nondinosaur footprints are *Rhynchosauroides* and *Batrachopus*.

New Jersey (Boonton Formation)

The Boonton Formation, youngest of the Newark Basin rock units, has yielded *Grallator, Anomoepus,* and *Batrachopus* tracks. Fossil fishes, including *Semionotus* and *Diplurus,* were discovered in this formation during the late 1800s.

Connecticut/Massachusetts (Shuttle Meadow Formation)

The earliest recorded appearance of dinosaurs in the Connecticut Valley is the footprints left on the ancient mudflats near Mt. Holyoke and Mt. Tom in Massachusetts. Among the theropod footprints from the Shuttle Meadow Formation of the Hartford Basin are *Eubrontes* and *Grallator.* A single tooth of an indeterminate small theropod was collected in 1970 from the Shuttle Meadow Formation in North Guilford, Connecticut, representing the oldest skeletal remains of a dinosaur yet found in the Connecticut Valley.

Connecticut/Massachusetts (East Berlin Formation)

The East Berlin Formation of the Hartford Basin has yielded many dinosaur tracks, but no dinosaur bones. Chief among the East Berlin footprint sites is Dinosaur State Park in Rocky Hill, Connecticut (chapter 9). Most of the Rocky Hill tracks are *Eubrontes*. At Westfield, Connecticut, *Grallator* tracks were discovered in this formation. The Westfield site produced the first fossil plants and fishes recorded in the Connecticut Valley (and perhaps in North America) in 1816. Also attributed to the East Berlin is the ornithischian footprint *Anomoepus*.

Connecticut/Massachusetts (Portland Formation)

The richest source in the Connecticut Valley for dinosaur footprints and bones is the Portland Formation, the youngest of the Hartford Basin strata. This formation supplied the famous brownstone from dozens of quarries throughout the Connecticut Valley.

The first recorded dinosaur tracks in North America came from the Portland Formation in 1802, when Pliny Moody uncovered birdlike footprints in South Hadley, Massachusetts (chapter 4). The Turners Falls, Massachusetts, area has produced more dinosaur tracks than any other region in the Connecticut Valley. Thousands of dinosaur prints came from sites like Horse Race, Lily Pond, and Field's Orchard near Gill, Massachusetts, and from Wethersfield Cove, Middletown, and Portland, Connecticut. These footprints include large and small *Grallator, Eubrontes,* and *Anomoepus,* plus the primitive crocodile-like track *Otozoum.*

The first recorded dinosaur bones in North America were found in 1818 in the Portland Formation at East Windsor, Connecticut. In addition, the greatest single source of dinosaur skeletons on the East Coast was the Buckland Quarry near Manchester, Connecticut (chapter 4). In the late 1800s this single brownstone quarry produced three skeletons of two new genera of prosauropod dinosaurs: *Anchisaurus* and *Ammosaurus.* Other major discoveries of dinosaur bones in the Portland Formation occurred in Massachusetts at Springfield in 1855 (*Anchisaurus*); South Hadley in 1910 (*Podokesaurus*); and Greenfield in 1875 (*Anchisaurus*). The Greenfield site is the northernmost locality for dinosaur bones yet discovered in the Connecticut Valley. Nondinosaur remains in the Portland include the skeleton of a foot-long, crocodile-like reptile called *Stegomosuchus* discovered in 1897 at Longmeadow, Massachusetts.

Nova Scotia (Scots Bay Formation)

The Scots Bay Formation of the Fundy Basin is a layer of gray and white limestone and brown sandstone that has yielded the theropod dinosaur tracks *Eubrontes* and *Grallator.*

Nova Scotia (McCoy Brook Formation)

The McCoy Brook Formation of the Fundy Basin is the lateral equivalent of the Scots Bay Formation. Sites near Parrsboro, Nova Scotia, have produced a wealth of Early Jurassic reptile fossils. Basalts from Wasson Bluff at Parrsboro are dated at approximately 200 million years old, thought to be right on the Triassic/Jurassic boundary. Dinosaur footprints found in the McCoy Brook Formation include *Anomoepus, Grallator,* and *Eubrontes.* The nondinosaurian tracks *Otozoum* and *Batrachopus* also appear. Dinosaur bony remains include skeletons of the prosauropods *Anchisaurus* and *Ammosaurus* and teeth from small theropods and ornithischians. Among the prize fossils from the McCoy Brook Formation are the remains of tritheledonts, close relatives to the oldest mammals.

TABLE 6.1   Early Jurassic Dinosaurs of the East Coast

Dinosaur	Locality	Geological Formation	Description[a]
Saurischia			
Podokesaurus holyokensis	Massachusetts	Portland Formation	S, bipedal ceratosaurian theropod, 1 m long. One of the best dinosaur skeletons from the Early Jurassic of the East Coast, now destroyed.
Anchisaurus polyzelus	Connecticut, Massachusetts, and ?Nova Scotia	Portland and McCoy Brook formations	S, prosauropod. 2.5 m long plant eater.
Ammosaurus major	Connecticut, and ?Nova Scotia	Portland and McCoy Brook formations	S, 4 m long prosauropod.
Grallator	Virginia, New Jersey, Connecticut, Massachusetts, and Nova Scotia	Turkey Run, Feltville, Towaco, Boonton, Shuttle Meadow, East Berlin, Portland, and McCoy Brook formations	F, small three-toed hind prints. Ceratosaurian theropod trackmaker 0.5 m at hips, 1.5 m long.
Eubrontes	Virginia, New Jersey, Connecticut, Massachusetts	Turkey Run, Towaco, Shuttle Meadow, East Berlin, and Portland formations	F, large three-toed hind prints. Ceratosaurian theropod trackmaker 5–6 m long, 1 m at hip.
Ornithischia			
Anomoepus	New Jersey, Connecticut, Massachusetts, and Nova Scotia	Feltville, Towaco, Boonton, Shuttle Meadow, East Berlin, Portland, and McCoy Brook formations	F, three-toed hind prints, four-toed fore prints. Quadrupedal ornithischian trackmaker about 1–2 m long, less than 1 m high at the hips.

[a]S = skeletal remains; F = footprints.

East Coast Dinosaurs of the Early Jurassic

The Early Jurassic dinosaurs of the East Coast are best known through the early efforts of Edward B. Hitchcock, Richard Swann Lull, Othniel C. Marsh, Edward D. Cope, and Mignon Talbot, and through the more recent fieldwork of researchers including Edwin H. Colbert, Donald Baird, John H. Ostrom, Paul E. Olsen, Walter P. Coombs Jr., Peter M. Galton, Hans-Dieter Sues, and Neil H. Shubin. Like those in the Late Triassic, the dinosaurs of the Early Jurassic left us more footprints than bones (table 6.1). Yet by combining information from all available fossil evidence we can learn what these East Coast animals were like, how they lived, and their place in evolutionary history.

Ornithischia: Tracks and Teeth Again

The first inklings of East Coast ornithischian dinosaurs were provided by the Late Triassic record of *Atreipus* footprints and a few rare teeth. This same theme—tracks and teeth—is all that is known of these small, bird-hipped plant eaters throughout the Early Jurassic of the Newark, Hartford, and Fundy basins of the Newark Supergroup (chapter 3).

From Nova Scotia in the north to New Jersey in the south, we find rare body fossils and abundant and widely distributed footprints of a new kind of primitive ornithischian. Called *Anomoepus* ("dissimilar foot"), this track was made by a 1 to 2 centimeter (0.6 to 0.8 inch) hind foot with three narrow and highly divergent toes and an occasional fourth sidetoe (fig. 6.1). It often includes a five-fingered handprint. At the Riker Hill Quarry near Roseland, New Jersey, there are even tiny *Anomoepus* (and *Grallator*) tracks that may have been made by baby dinosaurs.

Most *Anomoepus* tracks were made by animals walking or running. But some were apparently produced by a crouching or sitting trackmaker, for there is the distinct impression of the rest of the foot from the heel to the ankle joint. These same prints provide the impression of the first digit and give enough room for the fifth metatarsal, which apparently bore no toes. From this evidence we can reconstruct the digital formula of the hind foot of *Anomoepus*: 2-3-4-5-0. Likewise, the smaller forefoot is 2-3-4-3-2. These foot proportions and formulas compare well with the Late Triassic *Atreipus*.

In contrast to the footprints, the rare ornithischian skeletal remains are tantalizing at best and frustrating at worst. So far, all that has been discovered are yet to be described teeth, jaws, and other skeletal material, all from Nova Scotia. As with the Late Triassic ornithischians *Galtonia gibbidens* and *Pekinosaurus olseni*, comparisons with better-known relatives from elsewhere help us understand these Early Jurassic remains.

The teeth from Nova Scotia look similar to the cheek teeth of all primitive ornithischians: small, broad-based, triangular, and coarse along the front and

FIG. 6.1 Front- and hind-foot prints of *Anomoepus*, thought to have been made by a primitive ornithischian, from the Early Jurassic of the East Coast (after Olsen and Baird 1986).

back margins. These same features can be found in *Lesothosaurus* from southern Africa and in *Scutellosaurus*, a stegosaur-ankylosaur relative from the American Southwest. Both of these ornithischians were relatively small plant eaters, 1 to 1.5 meters (3.3 to 5 feet) long. They were also long-legged, bipedal runners, though *Scutellosaurus* may have rested and walked on all fours.

What picture emerges of the Early Jurassic ornithischians from the East Coast? We see the same four basic characteristics as in Late Triassic ornithischians (chapter 5), for the most part related to their plant-eating habits. These primitive ornithischians were small, no more than 2 meters (6.5 feet) long. The tail was probably long for balance, and the hind legs were longer than the forelegs. Because of the commonness of forefoot prints in *Anomoepus* tracks, it is likely that these animals walked on all fours but picked up their front limbs when running. Using both *Lesothosaurus* and *Scutellosaurus* for clues, the *Anomoepus* trackmaker may have sprinted at up to 5.5 meters a second (12.3 miles an hour). That compares to estimated speeds for walking on four legs of 1.4 meters a second (3.1 miles an hour).

The small, leaf-shaped teeth of these dinosaurs were probably adequate for simple chewing of succulent leaves and an occasional fruit or insect, but not anything like the complex chewing seen in many later ornithischian groups (chapters 1, 7, and 8). These animals foraged within the undergrowth of ferns, conifers, and ginkgoes, probably to a height of 1 meter (3.3 feet) or less, so they shared their food with low-browsing prosauropods and rhynchocephalians.

The Early Jurassic ornithischians did their best to escape from the jaws of the fierce carnivores whose tracks are mingled with theirs throughout the East Coast. Among the predators are the crocodile-like makers of *Otozoum* and *Batrachopus* tracks and the theropod maker of *Grallator* prints.

Yet speed and agility may not have been enough defense for large groups of ornithischians. The abundance of small *Anomoepus* tracks at New Jersey's Riker Hill provides strong support for gregarious behavior in dinosaurs, similar to that discussed for the Late Triassic trackmaker *Gregaripus* (chapter 5).

Theropoda: Skeletons, Fires, and Footprints

The fossil record for Early Jurassic theropod dinosaurs fits the same pattern as for the ornithischians: many footprints and trackways but few skeletal remains, with localities primarily in the Connecticut Valley, Newark Basin, and Fundy Basin.

The theropod bones include the mold of a partial hind leg and the fragmentary and now-destroyed skeleton of *Podokesaurus holyokensis* ("swift-footed reptile from Holyoke," named for Mount Holyoke College in Massachusetts). Fortunately, casts of *Podokesaurus* now reside at the Yale Peabody Museum and the American Museum of Natural History. This 1 meter (3.3 foot) long theropod has proved to be a puzzle, in part because of its incomplete nature, but also because the relationships of all small, meat-eating dinosaurs have been in great confusion.

When first discovered in 1910 in Massachusetts, *Podokesaurus* was known from part of the vertebral column (including some of the tail), a fragmentary humerus, some ribs, a pubis and ischium, a partial left femur, and much of the right leg (fig. 6.2). Although the bones were very poorly preserved, Mignon Talbot—who described this small dinosaur in 1911 (chapter 4)— noted features in the skeleton that suggested its uniqueness.

Three years later German paleontologist Friedrich von Huene (chapter 9) used *Podokesaurus* as the founding member of a new group of small theropods, Podokesauridae, under the umbrella of the so-called coelurosaurs. Despite such an exalted position, no additional evidence of *Podokesaurus* has ever been found. So the significance of this dinosaur came to depend on the original, poorly preserved skeleton.

Only twice has *Podokesaurus* had another look from dinosaur paleontologists. The first, by Richard Swann Lull of Yale University in 1915, built on and elaborated Talbot's earlier description. But it was the second look, a half century later, that tried to come to grips with this theropod. By that time, in 1964, important new and abundant material of the small theropod *Coelophysis* had been discovered in Upper Triassic rocks of New Mexico. Edwin H. Colbert (chapter 9) compared the surviving casts of *Podokesaurus* with the *Coelophysis* treasure trove, and his conclusion was that *Podokesaurus* and *Coelophysis* were the same kind of theropod. Because *Coelophysis* is the older name (dating to 1889), Colbert called the Massachusetts species *Coelophysis holyokensis*. There the matter rested until Olsen noted that the characters identified by Colbert as being uniquely shared by the two species are in fact found in many other theropods.

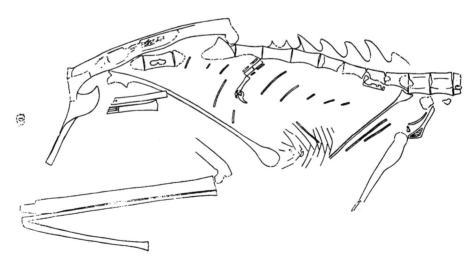

FIG. 6.2 *Podokesaurus,* an enigmatic theropod from the Early Jurassic of Massachusetts (after Talbot 1911).

So what exactly is *Podokesaurus?* It may be possible to identify this dinosaur better, largely because of recent research into the evolution of theropods (including birds, chapter 10). To begin with, *Podokesaurus* appears to lack cavities in the vertebrae of its trunk and has a short backbone relative to the length of the femur. These are features of the group known as tetanuran theropods, which includes *Tyrannosaurus, Deinonychus, Allosaurus,* and *Compsognathus.*

Is *Podokesaurus* a ceratosaur, one of that great group of theropods that shares closest relationship with these tetanurans? It may be, based on some liberal interpretations of the anatomy of the poorly preserved skeleton. The neck vertebrae found with the specimen exhibit one or more cavities—called *pleurocoels* ("side cavities")—that characterize all ceratosaurs. And if Lull's description of the flanges (transverse processes) that stick out laterally from each vertebra is correct, then their broad and swept-back condition is exactly what is seen in ceratosaurs.

So *Podokesaurus* may prove to be a ceratosaur, and if so it provides evidence—for the first time from skeletal remains—of a ceratosaur that lived on the East Coast during the Early Jurassic.

We can also glean some additional information about the biology of *Podokesaurus* from its skeletal proportions. This small theropod was fleet footed, simply because of the long and slim hind leg, in which the tibia (shinbone) and foot together are much longer than the femur. But we can do better than that. As in *Lesothosaurus,* relation of femur length to the rest of the leg can be used to calculate running speed. Thus *Podokesaurus* probably had a maximum running speed of 4 to 5.5 meters a second (9 to 12.3 miles an hour).

As in the Late Triassic on the East Coast, small and large theropod footprints known as *Grallator* appear at many sites, from Virginia to Nova Scotia. But joining *Grallator* in the Early Jurassic is a big newcomer whose tracks are called *Eubrontes* ("true thunder").

This contemporary of *Grallator* was much larger than previous theropod trackmakers (fig. 6.3). From the heel to the tip of the middle claw, *Eubrontes* tracks—given such species names as *Eubrontes approximatus* and *Eubrontes giganteus*—are typically 50 centimeters (20 inches) long. Originally named by Edward Hitchcock in 1845 for some of the footprints he was studying from Massachusetts, *Eubrontes* tracks appear to have a ?-3-4-5-? digital formula and often display impressions of the heel. Based on the size of these tracks, the *Eubrontes* trackmaker must have been over 1 meter (3.3 feet) high at the

FIG. 6.3 Hind-foot track of *Eubrontes*, thought to have been made by a large ceratosaurian theropod, from the Early Jurassic of the East Coast (after Lull 1953).

hip and 5 to 6 meters (16 to 20 feet) long, smaller but otherwise consistent with *Dilophosaurus* from the desert Southwest.

Because of its larger size and abundance in the Early Jurassic, *Eubrontes* is considered by most paleontologists to be a different track from *Grallator* and apparently represents the first large theropod on the East Coast. This new, big theropod was quite a runner, with speeds estimated at up to 4.5 meters a second (10 miles an hour). That's similar to the large individuals of *Grallator*, and attacks by both of these theropods would have been swift and violent for their unfortunate prey.

Eubrontes footprints also provide clues to theropods' swimming ability. At Dinosaur State Park in Rocky Hill, Connecticut (chapter 9), most *Eubrontes* prints show the usual three-toed impressions complete with digital pads and claw and heel marks. However, there are occasional prints and short trackways that show exceptionally clear claw marks, with very little toe impression and no heel whatever. In 1980 Walter P. Coombs Jr. (Western New England College, Springfield, Massachusetts) suggested that the unusual tracks were made by an animal swimming in shallow water, kicking the bottom with the tips of its toes. For track sequences that end abruptly, the theropod may have been buoyed up while swimming so it temporarily lost contact with the bottom.

There is also Early Jurassic footprint evidence that the makers of both *Grallator* and *Eubrontes* tracks were social and perhaps even hunted in packs. At

Mt. Tom in central Massachusetts, 134 prints were found arranged in twenty-eight trackways along a single bedding plane. About 70 percent of these tracks, mostly *Grallator,* are oriented in nearly parallel courses. John H. Ostrom of Yale University (chapter 9), who researched the Mt. Tom site, suggested that these theropods were traveling as a large group, all at the same time, across a broad mudflat.

At Dinosaur State Park, where most prints are *Eubrontes,* eighty-six trackways show preferred orientation, either to the northeast or southwest, but the pattern is not so striking as at Mt. Tom. To Ostrom, Rocky Hill trackways suggest that large theropods ambled across the mudflat over a much longer period than at Mt. Tom. In this way, Rocky Hill may represent several comings and goings of a few theropod herds.

Prosauropods: East Coast Browsing at Last

Prosauropods, the largest of Late Triassic and Early Jurassic herbivores, are among the most widespread of major dinosaur groups worldwide. And there is good evidence—skeletal remains for a change!—that prosauropods had finally entered the East Coast by the Early Jurassic, specifically in the Connecticut Valley (chapter 4).

The best-known East Coast prosauropod—*Anchisaurus polyzelus* ("near reptile, many rivals")—was named from skeletal material collected in 1855 near Springfield, Massachusetts. The dinosaur was first described as *Megadactylus polyzelus* in 1865 and later as *Amphisaurus* ("double reptile") *polyzelus* by Marsh. Both these names had already been used, so Marsh renamed the animal *Anchisaurus polyzelus* in 1885.

This was not the only evidence of *Anchisaurus* found in the Connecticut Valley. A nearly complete skeleton was discovered in the late 1800s in a brownstone quarry at Manchester, Connecticut, and much earlier, in 1818, several bones had been collected from East Windsor, Connecticut. In 1892 Marsh described these remains as belonging to another species of *Anchisaurus,* which he called *Anchisaurus colurus* (*colurus* means "short tail"). Not satisfied with this assessment, von Huene renamed this species *Yaleosaurus colurus* in 1932, in honor of Yale University. Most recently, Peter M. Galton regarded both *Anchisaurus polyzelus* and *Yaleosaurus colurus* as the same species, with the name *Anchisaurus polyzelus* having priority.

Anchisaurus was a relatively small—2.5 meters (8 feet) long—lightly built prosauropod, known from a nearly complete skull and skeleton missing only the tail and part of the neck. Its skull is small in proportion to the rest of the body. Even so, the snout is relatively long and slender, and the jaws are lined with teeth that bear coarse serrations on their front and back edges (fig. 6.4).

Like all prosauropods, *Anchisaurus* appears to have had a long and flexible neck. The trunk was somewhat rotund, supporting a roomy gut for fermenting plant food. The long, flexible tail helped balance the body over the hips. The hind legs were strong, and the narrow, four-toed foot was sturdy (the fifth toe is rudimentary in *Anchisaurus*). The shoulders and forelegs were also powerful. Within the forefoot, the outer fingers were small and slender and probably bore no claws. By contrast, the inner fingers—especially the first finger—were strongly built, the latter tipped by a greatly enlarged, sharply curved claw.

The other prosauropod from the Early Jurassic of the East Coast is *Ammosaurus major* ("greater sand reptile"), whose skeletal remains were found in the same Manchester quarry as *Anchisaurus*. Marsh originally named this dinosaur *Anchisaurus major* but renamed it *Ammosaurus major* in 1895. Also included within *Ammosaurus major* is the smaller *Anchisaurus solus,* another prosauropod skeleton found at the Manchester Quarry.

Ammosaurus is the broad-footed prosauropod of the Connecticut Valley. It is also reported from the Early Jurassic of Arizona and Nova Scotia. So far known from four incomplete juvenile and adult skeletons—unfortunately none with skulls—*Ammosaurus* grew to approximately 4 meters (13 feet) long. There are a number of unique pelvic modifications, but otherwise this dinosaur looks like many other prosauropods: long neck and tail, rotund body, stout hind legs, and strong forelegs.

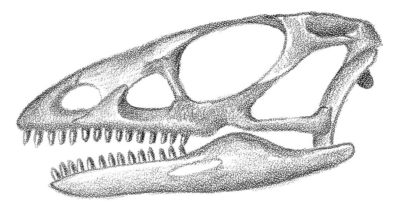

FIG. 6.4 Skull of *Anchisaurus,* a prosauropod from the Early Jurassic of the Connecticut Valley (after Galton 1976).

Despite their close proximity and nearness in time, *Anchisaurus* and *Ammosaurus* are not closely related: *Ammosaurus* is much further up the prosauropod evolutionary tree than *Anchisaurus*. But regardless of evolutionary relationship, both *Ammosaurus* and *Anchisaurus* were the dominant large terrestrial herbivores of the East Coast during the Early Jurassic.

There's a paradox here. Because of their lightly built skulls, slender teeth with minimal wear, and lack of evidence for cheeks, prosauropods were sometimes thought to be carnivorous scavengers or omnivores. Yet other characteristics of the skull and skeleton indicate that these dinosaurs were herbivores. For example, the jaw joint is slightly below the level of the upper and lower tooth rows. In this position the jaw muscles produce the powerful biting force needed for chewing fibrous plant tissues.

Further evidence for plant eating in prosauropods appears in the abdominal cavity. Here are occasionally found concentrated masses of polished stones called gastroliths (chapter 1). Their presence suggests that some prosauropods ripped apart plant food in the muscular gizzard just in front of the stomach. That gastroliths haven't been found with all prosauropod skeletons—including those of *Anchisaurus* and *Ammosaurus*—is a puzzle. Perhaps when the animals died their packet of gastroliths literally exploded with the rupture of the gas-filled innards. Or perhaps they were carried away from the decomposing skeletons by stream or river action.

However they processed their fodder, prosauropods were the highest browsers of the Early Jurassic. They were capable of rearing up on their hind legs to reach branches 3 to 4 meters (10 to 13 feet) above the ground. Would that have been easy for a prosauropod to do? That depends on whether they were bipedal or quadrupedal. Galton noted that all obviously quadrupedal dinosaurs had long trunks relative to the length of their hind legs, whereas bipedal dinosaurs had relatively short trunks. Prosauropods were midway between these two extremes. As a consequence, Galton regarded prosauropods as primarily quadrupedal but sometimes—as when running—adopting a bipedal stance. Among prosauropods, *Anchisaurus* is closest to purely bipedal dinosaurs, so perhaps it spent more time on its hind legs.

If footprints were available they certainly would provide better clues than body proportions to these questions of locomotion. Unfortunately, no prosauropod tracks and trackways have been found on the East Coast. This is truly a mystery, considering the abundance of footprints of theropods, ornithischians, phytosaurs, and other beasts. In the past some of these footprints from the Late Triassic and Early Jurassic were thought to have been

MIDDLE AND LATE JURASSIC: CRETACEOUS PRELUDE

It's sad to say, but the East Coast missed the heyday of the great Jurassic dinosaurs. In the two centuries of eastern dinosaur paleontology, not a trace has been found of the huge sauropods and ferocious theropods so well known from the Late Jurassic of western North America. With the last moments of sediment deposition in the Early Jurassic in Nova Scotia, New Jersey, and the Connecticut Valley, the dinosaur story for this geologic period on the East Coast was over.

Let's take a brief look at the dinosaurs that existed elsewhere during the 50 million year gap on the East Coast between the last Early Jurassic dinosaurs and the reappearance of dinosaurs in the Early Cretaceous.

Middle/Late Jurassic Dinosaurs: Giant Sauropods Rule

From the Middle Jurassic, about 178 to 157 Ma, the worldwide record of dinosaurs is at an all-time low. In the Southern Hemisphere there is a modest variety of the long-necked, long-tailed, large-bodied sauropods, including *Patagosaurus* ("Patagonian reptile") and *Lapparentosaurus* ("Lapparent reptile"), named for French paleontologist A. F. de Lapparent. But only a few theropods are at all well known, such as *Piatnitzkysaurus* ("Piatnitzky reptile"), named for Argentine geologist A. Piatnitzky.

In the Northern Hemisphere there is a bit more Middle Jurassic dinosaur action. Two Mexican sites have yielded footprints and skeletal material, and localities in England and France have given us additional sauropods and ankylosaurs. But we get the most complete picture from China. In Sichuan province complete skeletons have been discovered of the primitive stegosaur *Huayangosaurus* ("Huayang reptile"; Huayang is the old name for Sichuan Province), the small plant-eating ornithopod *Yandusaurus* ("Yandu reptile"; Yandu is the former name for Zigong municipality), and a host of gigantic sauropods. These include *Shunosaurus* ("Shuno reptile"; Shuno is the old name for the Sichuan region) and *Datousaurus* ("big chief reptile").

With the arrival of the Late Jurassic, many of these dinosaur groups positively bloomed. Never again would there be such a diversity of gigantic dinosaurs. And the smaller ones, both carnivores and herbivores, didn't do so badly either. The great discoveries in the western United States from the second half of the 1800s into the twen-

made by prosauropod dinosaurs, and they were given names like *Anchisauripus* and *Otozoum*. But when looked at in detail and compared with the fossil foot bones of prosauropods, it becomes clear that these "prosauropod" prints must have been made by theropods or nondinosaurian archosaurs. All we can say about the absence of prosauropod footprints in the East is that these dinosaurs apparently didn't frequent the mudflats surrounding the rift valley lakes.

tieth century provided us a clear vision of this great flowering of dinosaurs. Bones literally tumbled out of the ground at places like Como Bluff and Howe Quarry in Wyoming, Morrison and Cañon City in Colorado, and Utah's Cleveland-Lloyd Quarry and Carnegie Quarry (now Dinosaur National Monument).

These sites gave us many of the most familiar of all dinosaurs: fierce theropods like *Ceratosaurus* ("horned reptile") and *Allosaurus* ("strange reptile"); giant sauropods like *Brachiosaurus* ("arm reptile"), *Camarasaurus* ("chamber reptile"), *Diplodocus* ("double beam"), and *Apatosaurus* ("deceptive reptile"); the plated *Stegosaurus* ("roof reptile"); and the bipedal ornithopods *Dryosaurus* ("oak reptile") and *Camptosaurus* ("flexible reptile"). In Europe the Late Jurassic menagerie also included stegosaurs, ornithopods, theropods, and sauropods.

Farther to the east, China sports a diverse Late Jurassic dinosaur fauna that begins to rival that of the western United States. It includes the sauropods *Mamenchisaurus* ("Mamenchi reptile," named for Mamenchi Ferry) and *Omeisaurus* ("Omei reptile," named for the sacred mountain Mt. Omei); the theropods *Szechuanosaurus* ("Sichuan reptile") and *Yangchuanosaurus* ("Yangchuan reptile," named for Yangchuan County); and the stegosaurs *Tuojiangosaurus* ("Tuojiang reptile," named for the Tuojiang River) and *Chungkingosaurus* ("Chungking reptile," named for Chungqing City). Surprisingly, there were no ornithopods in Late Jurassic China.

In the Southern Hemisphere the biggest dinosaurian haul of all comes from the Tendaguru Hills in southern Tanzania, where an expedition from Berlin's Museum of Natural History collected hundreds of dinosaur skeletons from 1909 to 1912. This Tendaguru treasure trove included a small theropod called *Elaphrosaurus* ("fleet reptile"); the sauropods *Brachiosaurus, Dicraeosaurus* ("bifurcated reptile"), and *Barosaurus* ("heavy reptile"); the ornithopod *Dryosaurus;* and a spiky-backed stegosaur named *Kentrosaurus* ("sharp-pointed reptile").

These are the riches for which the East Coast has no record. The great dynasties of sauropods—the high-necked brachiosaurids, the majestically sleek diplodocids, and several other primitive forms—ruled all terrestrial habitats during the Late Jurassic. But they shared them with low-browsing stegosaurs and ornithopods and both small and enormous theropods, all agile and fearsome.

However, it is still possible to estimate walking and running speeds from skeletal remains, as we did with *Podokesaurus.* Both *Ammosaurus* and *Anchisaurus* may have taken leisurely quadrupedal walks at 0.5 to 0.8 meters a second (1.2 to 1.8 miles an hour). But at a bipedal full tilt they may have hit upward of 4 meters a second (9.3 miles an hour)!

While the animal was resting or walking on all fours, the enormous claw on the prosauropod thumb must have been held clear of the ground. But when these dinosaurs reared up or ran bipedally, the formidable claws were used for offense or defense. Against whom? Perhaps against each other, when engaged in territorial battles or defending mates. Or perhaps during mortal combat with predators: the likes of the *Eubrontes* trackmakers, packs of theropods leaving behind their *Grallator* tracks, and the makers of *Otozoum* prints. These were the life-and-death skirmishes of the Early Jurassic along the East Coast.

7

D I N O S A U R S
O F T H E   E A R L Y   C R E T A C E O U S

With the beginning of the Cretaceous, the sauropod-dominated dynasty of the Late Jurassic faded away in many regions of the world. On the East Coast of North America, the dinosaur story resumes after a 50 to 70 million year gap with a meager, but very important, Early Cretaceous fossil record from central Maryland and Washington, D.C. In fact it's the only known record of North American dinosaurs from this period east of the Mississippi River. Before we examine the East Coast record, let's take a look at Early Cretaceous dinosaurs elsewhere in the world.

Early Cretaceous Dinosaurs: Drifting and Differentiating

By the Early Cretaceous, fragmenting and shifting landmasses put a stop to long-distance interactions among terrestrial animals. The dinosaurs became much more differentiated in both the Southern and Northern Hemispheres. In South America we find new and often bizarre theropods and sauropods, but the best-preserved of the southern remains come from sub-Saharan Africa and eastern Australia. Between the two, there are important new ornithopods, theropods, sauropods, and ankylosaurs.

In more northerly latitudes, from eastern Asia to the western United

States, we find a host of new dinosaur groups. Among them were the first ceratopsians, new kinds of ankylosaurs (called ankylosaurids), and the theropods known as ornithomimosaurs, troodontids, and dromaeosaurids.

Much of this variety is seen in Mongolia and China, where the 2 meter (6 foot) long *Psittacosaurus*, a primitive ceratopsian, foraged through the Early Cretaceous shrubbery with the ankylosaurid *Shamosaurus* ("desert reptile"), the stegosaur *Wuerhosaurus* ("Wuerho reptile," named for Wuerho, China), the large plant-eating ornithopod *Probactrosaurus* ("before *Bactrosaurus*"), the primitive ornithomimosaur *Harpymimus* ("Harpy mimic"), duck-billed hadrosaurids, and several kinds of sauropods. There were also skeletal remains of large and small theropods, and footprints of both theropods and ornithopods.

The dinosaur population in Europe was large and diverse, with the best record found in western Europe. Here there are footprints and skeletal fossils galore, including two of the three charter members of Sir Richard Owen's Dinosauria: the ornithopod *Iguanodon* and the ankylosaur *Hylaeosaurus*. New theropods from Europe include the recently discovered *Baryonyx* ("heavy claw"), while the giant sauropods include *Aragosaurus* ("Aragon reptile," named for the Aragon region of Spain) and *Pelorosaurus* ("colossal reptile"). The ankylosaur *Acanthopholis* ("spine scute") fed on ground cover with such small ornithopods as *Hypsilophodon* and *Valdosaurus*.

Europe was also home to the first known thick-headed dinosaurs, or pachycephalosaurs, such as *Yaverlandia*. In addition, the region boasts the earliest occurrence of high-latitude dinosaurs. In far-off Svalbard—a lonely, freezing island better known as Spitsbergen, off the northern coast of Norway—footprints of both theropods and ornithopods reveal the presence of dinosaurs above the Early Cretaceous Arctic Circle.

Finally, we reach the Early Cretaceous in North America. In the western and southwestern United States, the rich assortment of sauropods and stegosaurs of earlier times had diminished in number and diversity. In their place were a host of ornithopods, among them *Zephyrosaurus* ("west wind lizard") and *Tenontosaurus* ("sinew reptile"), plus ankylosaurs like *Sauropelta* ("shield reptile") and *Nodosaurus* ("knob reptile") and a few hadrosaurids. These plant eaters formed the tasty repast for many contemporary predators like *Microvenator* ("small hunter") and *Deinonychus* ("terrible claw"), the model for *Jurassic Park*'s blood-chilling raptors.

The Midwest of the United States also provides a rare glimpse of Early Cretaceous dinosaurs, including ankylosaurs like *Silvisaurus* ("forest reptile"), or-

nithopods like *Tenontosaurus,* theropods like *Acrocanthosaurus* ("high-spined
reptile"), and a few sauropods.

But it's their abundant footprints that tell us dinosaurs were commonplace
in the North American Midwest during the Early Cretaceous. In Oklahoma,
Texas, and Arkansas there are prints of theropods, ornithopods, and
sauropods. Many of these are preserved in trackways that show not only the
pattern of footfall of each of these kinds of dinosaurs, but also their rate of
progress and even aspects of tracking and attacking.

East Coast Dinosaur Distribution

As we learned in chapter 3, the Maryland and District of Columbia portion
of the Early Cretaceous Atlantic Coastal Plain consists of a stack of sediments
laid down in large rivers or along deltas that opened into the surrounding
shallow seas to the east. Waters flowed sluggishly through these environments, now represented by the Patuxent and Patapsco Formations. But it is

TABLE 7.1   Early Cretaceous Dinosaurs of the East Coast

Dinosaur	Locality	Geologic Formation	Description[a]
Saurischia			
"Ornithomimus" affinis	Maryland	Arundel Clay	S, small ornithomimosaur, 2 m long.
Indeterminate theropod	Maryland and Washington, D.C.	Arundel Clay	S, mixture of fragmentary bones and teeth of meat-eating dinosaurs.
Astrodon johnstoni	Maryland and Washington, D.C.	Arundel Clay	S, 20 m long sauropod known from both juve-nile and adult material. Quadrupedal plant eater.
Ornithischia			
Priconodon crassus	Maryland	Arundel Clay	S, 5–6 m long armored nodosaurid ankylosaur. Quadrupedal plant eater.
?Tenontosaurus sp.	Maryland	Arundel Clay	S, 7 m long ornithopod. Bipedal plant eater.

[a]S = skeletal remains; F = footprints.

in the intermediate Arundel Clay, not the Patuxent or Patapsco, that the only Early Cretaceous dinosaurs along the East Coast have been found (table 7.1).

Maryland/Washington, D.C. (Arundel Clay)

Fossil evidence for the first sauropod dinosaur in North America was discovered in 1859 in Bladensburg, Maryland: two teeth named *Astrodon*. Sites at Muirkirk, Maryland, and elsewhere in the Arundel Clay between Baltimore and Washington, D.C., have produced additional skeletal remains of *Astrodon* (formerly known as *Pleurocoelus*), several indeterminate theropods, a tenontosaur, and the ankylosaur *Priconodon*. More evidence of Arundel dinosaurs was found in Washington in 1898 (an indeterminate theropod) and 1942 (*Astrodon*).

East Coast Dinosaurs of the Early Cretaceous

The East Coast fossil record of Early Cretaceous dinosaurs was collected or studied early on by Othniel C. Marsh, John Bell Hatcher, Arthur B. Bibbins, Richard Swann Lull, and Charles W. Gilmore, and in recent times by Peter M. Kranz, Michael K. Brett-Surman, Charles S. Martin, and others.

The Arundel fauna might be considered poor because it is a mixture of isolated bones of numerous kinds of dinosaurs. Obviously it would be better to have complete or even partial skeletons. But this East Coast record presents an important picture of the transition from Jurassic to Cretaceous dinosaurs. In addition, it boasts nearly a full complement of the various Early Cretaceous dinosaur groups found elsewhere in North America and the world.

Ankylosauria: Hunkering down under Armor

The Arundel ankylosaur *Priconodon crassus* ("thick saw-cone tooth") is known only from isolated teeth, first named and described by Marsh in 1888. This animal is a member of Nodosauridae, one of the two great groups of ankylosaurs (the other is Ankylosauridae). Nodosaurid affinities for *Priconodon* are based on the large size of the teeth, the narrowness of the tooth crown, and the presence of a shoulderlike shelf between the base and the crown (fig. 7.1).

Although no other *Priconodon* skeletal remains have been found on the East Coast, we can still gain insight into the anatomy and behavior of this Arundel dinosaur by looking at some of its nodosaurid relatives, among them *Sauropelta*, *Silvisaurus*, and *Panoplosaurus* ("fully armored reptile") from western North America.

FIG. 7.1 A tooth of *Priconodon,* a nodosaurid ankylosaur from the Early Cretaceous of Maryland (after Lull 1911b).

All nodosaurids were covered with bony armor, from the top of the head to the tip of the tail. With bodies 5 to 6 meters (16 to 20 feet) long, these animals were clearly outfitted for protection. Although nodosaurids lacked the famous tail club found only in ankylosaurids, they did have long, sharp spines over the shoulders that could have made large gashes in attackers. With its legs folded under its body, a multiton nodosaurid would have been very difficult to flip over. Safe under its protective armor, hunkered down to wait out the assault, *Priconodon* was virtually impregnable to predators.

Built sturdy and close to the ground, nodosaurs like *Priconodon* foraged on shrubby plants, ground cover, or leaves and fruits of the lower branches of Early Cretaceous trees. Such plants included low-growing ferns, cycads, and the earliest of flowering plants. These dinosaurs had a relatively narrow, scoop-shaped beak for snapping up vegetation. Morsels of plant food were sliced and diced by the teeth and further processed for digestion through gut fermentation. In all ankylosaurs, there is plenty of room beneath the very deep ribcage and wide pelvis to accommodate a capacious hindgut.

Priconodon and other ankylosaurs were near the bottom of the dinosaur ladder in terms of brains and speed. In general these armored dinosaurs had smaller brains for their body size than any other dinosaurs except the gigantic sauropods. Ankylosaurs were also among the slowest of all dinosaurs, capable of running no more than 2.7 meters a second (6 miles an hour). They usually moved about at a much more leisurely pace, approximately 1 meter a second (2.2 miles an hour).

Ornithopoda: Teeth Tell the Tale

The Arundel Clay has yielded a very meager record of ornithopods. So far the only discoveries are a large, fragmentary tooth that may have come from

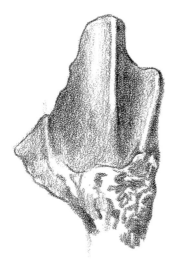

FIG. 7.2 A euornithopod tooth from the Early Cretaceous of Maryland (after Galton and Jensen 1979).

an animal like *Tenontosaurus* and a recently found smaller tooth that is still unnamed.

The large, fragmentary specimen consists of a broken crown and part of the root of a tooth from the left lower jaw (fig. 7.2). This tooth displays features that are characteristic of euornithopods, an ornithopod group (chapter 1) that includes hypsilophodontids and iguanodontians. The tooth was referred in 1978 by Peter M. Galton (University of Bridgeport) and James A. Jensen (then at Brigham Young University) to *Tenontosaurus*, a euornithopod from the Early Cretaceous of the Western Interior of the United States.

In contrast the small, unnamed tooth, found in 1990 by Peter M. Kranz (chapter 9), consists of the entire crown and part of the adjacent root of a tooth from the left lower jaw. Still under study, this tooth is smaller than one would expect of a full-grown tenontosaur, so it might have come from a juvenile. There is also a possibility the tooth belonged to a primitive ceratopsian dinosaur. This would be a spectacular discovery, for ceratopsians are known only from western North America and central and eastern Asia.

If the large tooth came from a tenontosaur or another primitive euornithopod, then we can take our clues from the better-known *Tenontosaurus* to flesh out the details of anatomy and ecology of these Early Cretaceous herbivores that lived in the Maryland/Washington, D.C., area some 115 Ma.

Based on their size and limb anatomy, ornithopods like *Tenontosaurus* were predominantly bipedal, with exceptionally long and quite muscular tails. Such features enabled these animals to hold their tails almost horizontally for balance. *Tenontosaurus* might have been able to travel at up to 4 to 5.5 meters a second (9 to 12.3 miles an hour) during a sustained bipedal run.

An animal the size of *Tenontosaurus* probably foraged close to the ground but was also able to rear up on its hind legs to nip at leaves some 2 meters (6.6 feet) above the ground. The front of the narrow beak lacked teeth and was covered with a sharp, horny covering. Leaves were probably brought into the mouth by a long, fleshy tongue, and these gulps of food were then intently chewed. For tenontosaurs and other euornithopods, chewing appears to have involved pleurokinesis ("side mobility"), a slight rotation of portions of the upper jaw. This was an important advance for euornithopod dinosaurs: it gave them the ability to chew a variety of plant foods, including those with a great deal of fiber.

Like the ankylosaur *Priconodon,* the Arundel tenontosaurs probably fed on conifers, deciduous shrubs, and trees of the newly evolved angiosperms. But unlike *Priconodon,* the ornithopods could use their free hands to bring leaves and branches closer to the mouth for more efficient feeding.

Sauropoda: First Giants of the East Coast

The first dinosaur discovered from the Maryland/Washington, D.C., area—and the second discovered on the East Coast—was a sauropod named *Astrodon johnstoni.* It was in 1859, a year after the christening of New Jersey's *Hadrosaurus foulkii,* that Dr. Christopher Johnston, a local physician, described a sauropod tooth (fig. 7.3) from Bladensburg, Maryland (chapter 4). Johnston called this tooth *Astrodon* ("star tooth") for its star-shaped appearance in cross section. In 1865 Joseph Leidy gave it the official species name *Astrodon johnstoni.*

Thereafter, additional sauropod remains began turning up from the Arundel Clay, with much of the material given the name *Pleurocoelus* ("side cavities"). In 1888 Marsh recognized the species *Pleurocoelus nanus* (*nanus* means

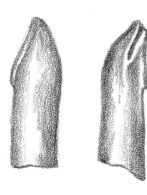

FIG. 7.3 A tooth of *Astrodon,* a brachiosaurid sauropod from the Early Cretaceous of Maryland (after Leidy 1865).

"dwarf") from skull material and isolated remains of more than six individuals. Marsh also described another species—*Pleurocoelus altus* (*altus* means "old")—based on a tibia and fibula (lower hind-leg bones). A year later, Richard Lydekker of London's British Natural History Museum described a new sauropod from the Early Cretaceous of southeastern England as *Pleurocoelus valdensis* (*valdensis* means "from the Weald"). In 1974 Wann Langston Jr. identified *Pleurocoelus* material from Texas.

Although three names were historically given to the material from the Arundel Clay of Maryland and Washington, D.C., it appears there is really only one sauropod, represented by both young and old individuals. Some of the vertebrae of this long-necked dinosaur have huge chambers (called *pleurocoels*) within them, larger than in any other sauropod. Most of the remains are small by sauropod standards, probably belonging to juvenile animals that weighed no more than 500 kilograms (1,000 pounds) and measured less than 5 meters (15 feet) in length. A few bones suggest animals of more typical sauropod size. These were probably adult individuals that were up to 18,000 kilograms (20 tons) and 20 meters (66 feet).

Despite the isolated and often fragmentary preservation of the skeletal remains, we can still provide this dinosaur's proper name and place in the evolutionary context of Sauropoda and the Early Cretaceous world. Clearly, a single kind of sauropod can't have three names. Of the three, *Astrodon johnstoni* is based on an isolated tooth, whereas the two species of *Pleurocoelus* are based on much more material. The *Astrodon* tooth has several unique characteristics that are shared by only one other known sauropod: *Pleurocoelus nanus.* With features just like *Astrodon,* these *Pleurocoelus* teeth must belong to our lone Arundel sauropod. And because the name *Astrodon* was coined before *Pleurocoelus, Astrodon* has priority in any disputes about naming.

What kind of sauropod was *Astrodon?* The closest comparison is with the group known as Brachiosauridae (fig. 7.4), which has been extensively studied by sauropod specialist John S. McIntosh (retired, Wesleyan College). *Astrodon* possesses at least four of the distinctive characteristics of this group, including long forelimbs and hands, plus pleurocoels in the vertebrae in front of the pelvis. That makes *Astrodon* a brachiosaurid, one of the few Early Cretaceous survivors of this otherwise Jurassic group.

Combining the anatomy of *Astrodon* and other brachiosaurids, it may be possible to glean something of the behavior of the Arundel sauropod. The first item on the agenda is getting *Astrodon* out of the swamps. True, this is the

FIG. 7.4 *Brachiosaurus,* the best known of brachiosaurid sauropods, from the Late Jurassic of the western United States and Tanzania.

depositional environment of the Arundel Clay, where *Astrodon* bones and teeth have been found. But these remains were found widely separated, as you'd expect if an animal lived and died in another habitat, its remains pulled apart and scattered as the carcass traveled downstream.

More important, the once popular idea that sauropods frequented lakes, rivers, and swamps to buoy up their great bulk has been discredited. All sauropods, including *Astrodon,* had elephantine legs and feet (fig. 7.5) that were strong enough for walking on land at a slow pace. These giants also had

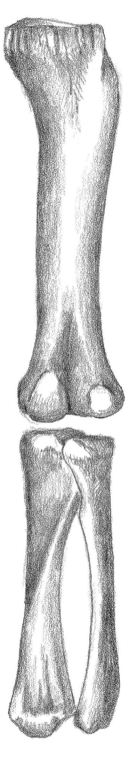

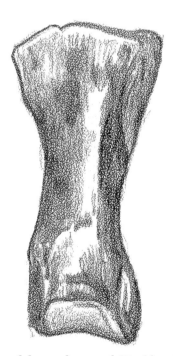

FIG. 7.5 Hind-leg bones of *Astrodon* (after Lull 1911b).

FIG. 7.6 One of the toe bones of *"Ornithomimus" affinis,* an ornithomimosaurian theropod from the Early Cretaceous of Maryland (after Gilmore 1921).

a narrow, slab-sided thorax, very different from the round ribcage of the amphibious hippopotamus but similar to terrestrial rhinoceroses and ele-
phants.

But what were the long necks for, if not for keeping the head above water? Ask a giraffe! Long necks in terrestrial animals are for feeding in tall trees.
With a long neck and long front legs, *Astrodon* could raise its head 10 meters
(33 feet) or more above the ground to feed on foliage that virtually no other
animal could reach. These plants included a great variety of conifers, ginkgoes, and ferns. Angiosperms were still a small part of the menu, since they
had not yet reached tall-tree height.

Surprisingly, brachiosaurid skulls and teeth don't appear to be well designed for chewing. Yet the *Astrodon* teeth from the Arundel Clay often display visible wear at their tips. Perhaps this wear came from stripping off foliage from high branches and delivering this mouthful to the gullet without
much chewing. As we learned in chapter 6, plant food was ground up for further digestion in the muscular gizzard.

We know a great deal about sauropod locomotion from both skeletons and
trackways. It appears that *Brachiosaurus,* and probably *Astrodon,* could run up
to 8 meters a second (18 miles an hour). But most of the time these behe-
moths probably walked a good deal slower. Sauropod trackways from elsewhere in the world also hint at some fairly complex social behavior. These
house-sized animals may have engaged in herding, migration, and even
parental care. Such behavior may have been a consequence of the age-old
problem of protection against predators.

Theropoda: East Coast Ostrich Mimics

The Arundel theropods are represented by a number of poorly preserved
teeth, vertebrae, and limb bones which have been named and renamed over
the years. *Allosaurus medius* ("middle strange reptile"), named by Marsh in
1888, is based on a single tooth. At the same time, Marsh described a new
theropod, *Coelurus gracilis* ("slender hollow tail"), from an isolated hand claw.
Creosaurus potens ("powerful flesh reptile"), named by Lull in 1911, is based
on a single tail vertebra.

In 1920 the Smithsonian's Charles W. Gilmore (chapter 9) referred these
theropods to the genus *Dryptosaurus,* otherwise known from the Late Cretaceous of New Jersey. Gilmore also described a new theropod, *Ornithomimus
affinis* ("related to the bird mimic"), from a recurved foot claw, two small tail
vertebrae, another partial vertebra, an anklebone, and two incomplete

metatarsal bones (fig. 7.6). These remains had been mistakenly referred by Lull to an ornithopod that he named *Dryosaurus grandis* ("large oak reptile").

Lots of other theropod specimens have been discovered over the years, and these have occasionally been referred to one or the other of these theropods. In 1972 Dale A. Russell (National Museum of Nature in Ottawa, Canada) undertook to review all North American ornithomimids, as well as a few Asian forms. Russell reassigned *Ornithomimus affinis* to his new genus *Archaeornithomimus* ("ancient bird mimic"), making it *Archaeornithomimus affinis*. In 1990 Peter Galton and David Smith (University of Bridgeport) took another look at *Archaeornithomimus*. They argued that *Archaeornithomimus affinis* is not particularly persuasive as an ornithomimid and thus described it as a poorly preserved small theropod.

As for *Creosaurus potens, Allosaurus medius,* and *Coelurus gracilis,* many dinosaur paleontologists have given thumbs down to their legitimacy. In all cases these theropods are now considered indeterminate as to genera and species.

That's where the Arundel theropods stood until 1992, when important new material was discovered. This specimen—the top of a left tibia (fig. 7.7) and a number of toe bones (including one of the claws)—probably came from an animal 3 to 4 meters (10 to 13 feet) long. It may be that we can narrow down the identity of this new theropod as well as that of *Ornithomimus affinis*. Among theropods, the only ones that have flat claws on the feet and a

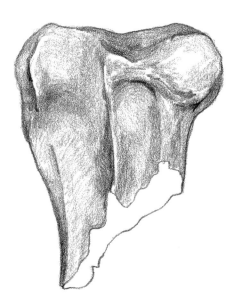

FIG. 7.7 Top of the tibia (shinbone) of an undescribed ornithomimosaurian theropod from the Early Cretaceous of Maryland.

FIG. 7.8 The ornithomimosaur *Dromiceiomimus*, from the Late Cretaceous of Alberta, Canada.

tibia that has a large crest sticking out the front are ornithomimosaurs. These are the same features found in the new Arundel theropod and in *Ornithomimus affinis*, so it is reasonable to conclude that they too are ornithomimosaurs.

Ornithomimosaurs were slender animals 3 to 5 meters (10 to 16 feet) long, clearly built for running (fig. 7.8). They also had long forelimbs and resembled modern ground-dwelling birds with their beaklike, toothless jaws. Among the fleetest of dinosaurs, these theropods ran at up to 16 meters a second (36 miles an hour). They were also relatively large brained, similar to ostriches and emus. Eyesight appears to have been keen and binocular, and their hearing was probably acute. Without teeth, ornithomimosaurs likely fed on soft foods such as insects, larvae, small vertebrates, eggs, and fruits. The strong and mobile forelimbs may have been used to shake food down from branches and manipulate it to the mouth.

Although *Ornithomimus affinis* seems to fit the description of an ornithomimosaurian theropod, it is probably inappropriate to call it *Ornithomimus,* or even *Archaeornithomimus,* since it likely isn't a close relative of the true species of either. All we can say at present is that there appears to be an ornithomimosaur in the Arundel fauna that perhaps should be called *"Ornithomimus" affinis,* a less than definitive answer that also applies to the new partial hind limb. Are there two different ornithomimosaurs? Only with additional, better-preserved material can we tell.

What about all those theropod teeth from the Arundel Clay? They belong not to ornithomimosaurs, but to the more conventional variety of toothed theropods. Their identity remains a mystery, as does the identity of their victims. It may be that these East Coast predators fed on tenontosaurs or *Astrodon,* at least young and feeble individuals. Their diet may also have included the fleet Arundel ornithomimosaurs, if they could catch them.

End of the Early Cretaceous

With the last of the dinosaurs of the Arundel fauna, we come to the close of the Early Cretaceous record of these animals along the East Coast. It is still an active and exciting arena for collecting and research, not least because of the relationship between these dinosaurs and the newly evolving angiosperms, the flowering plants that soon came to dominate the world's floras.

By the time we next encounter East Coast dinosaurs—in the Late Cretaceous—these animals were sharing their environments with a great abundance of angiosperms of all sorts, from shrubs to tall trees. It was a time of great beginnings, and endings.

8

DINOSAURS
OF THE LATE CRETACEOUS

The beaches, deltas, and hills of the eastern shores of North America teemed with dinosaur life in the lush Late Cretaceous. There were 10 meter (33 foot) long duck-billed dinosaurs, large and small theropods, armored ankylosaurs—all foraging for food, caring for their young, stalking and killing, defending against predators.

And dying. An occasional gas-bloated carcass drifted downriver, moving slowly to the mouth of a delta and then out to sea. If mosasaurs and crocodiles didn't make a meal of the decaying body, other ocean-dwelling animals had their feast. Fishes nibbled at the slowly rotting flesh. Fingers and toes, teeth, bits of the head and parts of the neck came loose. Eventually the voyage of our dinosaur ended as its remaining skeleton sank to a muddy ocean bottom tomb.

So it was with the Late Cretaceous dinosaurs of the East Coast. Although the western coastal plain rocks from this time period are gone, we have a great mass of sediments deposited in nearshore and deeper marine waters that lay immediately to the east of dry land (chapter 3). And it is to these marine beds that we look for our dinosaur record.

Before we tackle these last dinosaurs of the East Coast, what was happening elsewhere? During the Late Cretaceous, these Mesozoic beasts were more abundant, widely distributed, and diverse than ever before. From southern Argentina to the North Slope of Alaska, from Japan to California, it was a dinosaurs' paradise.

Late Cretaceous: Last of the Best

The continent of South America, fully separated from the African landmass by the Late Cretaceous, was home to a peculiar array of dinosaurs, including many new kinds of both large and small theropods and even a few duck-billed hadrosaurids. Across the ever-expanding South Atlantic Ocean in Africa, the record of dinosaurs so far consists of a few theropod footprints, plus the skeletal remains of some new sauropods, the only pachycephalosaur yet known from the Southern Hemisphere—*Majungatholus* ("Majunga dome"), discovered in Madagascar—and a variety of theropod carryovers from the Early Cretaceous.

India, not yet part of the Northern Hemisphere, was home to a menagerie of unusual theropods, some sauropods, and the last surviving stegosaur, *Dravidosaurus* ("Dravidanadu reptile," named for Dravidanadu, India). Australia and New Zealand, still firmly attached to Antarctica in the Late Cretaceous, have yielded a few dinosaur fossils. And ice-covered Antarctica had been barren ground for dinosaur hunters until a team of scientists from Argentina uncovered new ankylosaur material from Seymour Island, just south of Tierra del Fuego, in 1986.

By contrast, the continents of the Northern Hemisphere provide the best record of the rich dinosaur diversity of the Late Cretaceous. From Asia—particularly the rich beds of the Gobi Desert in Mongolia and adjacent China—come many fierce and strange-looking theropods. These include the gigantic *Tarbosaurus* ("terror reptile"), the small, aggressive *Velociraptor* ("swift stealer"), the truly bizarre *Oviraptor* ("egg stealer"), and the ostrichlike *Gallimimus* ("chicken mimic"). Duck-billed dinosaurs are represented by *Saurolophus* ("ridged reptile") and *Shantungosaurus* ("reptile from Shandong Province"). There is also a rich array of ankylosaurs, primitive ceratopsians like *Protoceratops* ("first horned face") and *Bagaceratops* ("small horned face"), and a number of flat- and dome-headed pachycephalosaurs.

Recently a brand new group of dinosaurs was discovered in Mongolia. Called segnosaurs (chapter 1), these peculiar saurischians look like a cross between a prosauropod, an ornithopod, and a theropod! There are even a few sauropods from these rich central Asian fossil fields, among them *Ne-*

megtosaurus ("reptile from Nemegt") and *Opisthocoelicaudia* ("hollow-backed tail").

Europe was home to hadrosaurids, ankylosaurs, ornithopods, and a sauropod or two. These European sites were originally part of a chain of islands that extended from Romania and Ukraine in the east to Spain and France in the west, and the dinosaurs show a number of peculiar features that attest to their island habits.

This abundance and diversity of dinosaurs is eclipsed by what has been found in the Western Interior of the United States, Canada, and Mexico. From numerous localities—particularly from Alberta, Canada, and adjacent Montana—we have a great variety of theropods, including ornithomimosaurs, small and agile predators like *Dromaeosaurus* ("swift reptile"), *Troodon* ("wounding tooth"), and of course the famous *Tyrannosaurus* ("tyrant reptile").

And then there are the dinosaurs that served as prey for these carnivores. The ornithopods include *Orodromeus* ("mountain runner") and duckbills like *Maiasaura* ("good mother reptile"), *Saurolophus, Lambeosaurus* ("Lambe reptile," named for Canadian paleontologist L. M. Lambe), and *Parasaurolophus* ("like *Saurolophus*").

In addition to ornithopods, there are more ceratopsians in western North America than have been found anywhere else in the world, among them *Triceratops* ("three-horned face"), *Centrosaurus* ("spur reptile"), and *Chasmosaurus* ("opening reptile"). This broad North American region also ranks neck-and-neck with central Asia for diversity in pachycephalosaurs and ankylosaurs. The pachycephalosaurs include *Stegoceras* ("horned roof") and *Pachycephalosaurus* ("thick-headed reptile") itself. The ankylosaurs are represented by *Euoplocephalus* ("well-armored head") and *Panoplosaurus* ("well-armored reptile").

Why have we found so many dinosaurs from the Late Cretaceous in western North America? In addition to the great exposures of Upper Cretaceous rocks in the West, the most compelling reason for such richness is that the area was an extensive and broad coastal plain inhabited by many kinds of dinosaurs. It was also a very dynamic coastal plain, cut by rivers that carried large amounts of mud and sand from the newly forming Rocky Mountains to the west. It shouldn't be a surprise that some of these dinosaurs were entombed and eventually made their way into the hands of paleontologists.

Elsewhere in North America, the American Southwest adds to the Late

TABLE 8.1   Late Cretaceous Dinosaurs of the East Coast

Dinosaur	Locality	Geologic Formation	Description[a]
Saurischia			
Dryptosaurus aquilunguis	New Jersey and North Carolina	Marshalltown, Mount Laurel/Wenonah, Navesink, and Black Creek formations	S, large, 4–5 m long bipedal theropod. Forelimbs bear an enormous claw.
Coelosaurus antiquus	New Jersey, Delaware, and Maryland	Marshalltown, Navesink, and Severn formations	S, 2 m long ornithomimosaur.
Diplotomodon horrificus	New Jersey	Navesink Formation	S, a single tooth of a theropod dinosaur.
Indeterminate ornithomimosaur	New Jersey and Delaware	Merchantville and Mount Laurel/Wenonah formations	S, a mixture of ornithomimosaurian theropod material.
Indeterminate theropod	New Jersey	Raritan and Marshalltown formations	F, S, three three-toed hind prints. Large theropod printmaker. Mixture of skeletal material from large theropods.
Ornithischia			
Indeterminate nodosaurid	New Jersey	Navesink, Mount Laurel/Wenonah, and Marshalltown formations	S, 5–6 m long armored ankylosaurian dinosaur.
Hadrosaurus foulkii	New Jersey	Woodbury, Merchantville, and Marshalltown formations	S, 10 m long duck-billed dinosaur. Bipedal plant eater. The most complete dinosaur skeleton from the Late Cretaceous of the East Coast.
"Hadrosaurus" minor	New Jersey	Navesink Formation	S, 8–10 m long duck-billed dinosaur. Bipedal plant eater.
Hypsibema crassicauda	North Carolina	Black Creek Formation	S, 12 m long duck-billed dinosaur. Bipedal plant eater.
Lophorhothon	North Carolina	Black Creek Formation	S, small duck-billed dinosaur. Bipedal plant eater.
Indeterminate hadrosaurid	New Jersey, Delaware, Maryland, North Carolina, South Carolina	Woodbury, Merchantville, Mount Laurel/Wenonah, Marshalltown Navesink, Severn, and Black Creek formations	S, mixture of bones and teeth of duck-billed dinosaurs, many of which were 10 m long. All were bipedal plant eaters.

[a]S = skeletal remains; F = footprints.

Cretaceous sauropod record with *Alamosaurus* ("Alamo reptile," named for the Ojo Alamo Formation of New Mexico). And from the Gulf Coast region of the United States—Missouri, Mississippi, Alabama, and adjacent parts of Georgia—there is a meager record from marine beds of hadrosaurids like the small-crested *Lophorhothon* ("crest nose"), some theropods, and even a few ankylosaurs.

East Coast Dinosaur Distribution

The band of Upper Cretaceous marine rocks that outcrops along the Gulf Coast continues east and northward along the eastern seaboard from Georgia to New Jersey. Dinosaurs have been found throughout these formations from as far south as central South Carolina to north-central New Jersey (table 8.1). This record, though patchy, spans the entire Late Cretaceous from 97 Ma to 65 Ma.

South Carolina (Donoho Creek Formation)

Several sites in the Donoho Creek Formation (formerly the Black Creek Formation) in northeastern and east-central South Carolina have yielded indeterminate dinosaur remains (chapter 9). A tooth and bone fragments of a theropod came from a site along the Peedee River in Florence County. The Quinby area, also in Florence County, has produced theropod and hadrosaurid teeth and a theropod foot-bone fragment. Another theropod tooth was found at a site near Lynchburg on the Lynches River. In addition, two hadrosaurid teeth came from Kingstree in Williamsburg County, although they were mixed with Paleocene and Pleistocene fossils and their stratigraphic source is uncertain.

North Carolina (Black Creek Formation)

Although the Upper Cretaceous Black Creek Formation sediments that have yielded the only dinosaur fossils in North Carolina are now properly known as the Donoho Creek Formation, the historical designation is still widely used. The most productive Black Creek site in North Carolina is at Phoebus Landing on the Cape Fear River (chapter 9). Among the discoveries at Phoebus Landing are the hadrosaurids *Hypsibema* and *Lophorhothon*, the theropods *Coelosaurus* and *Dryptosaurus*, indeterminate hadrosaurids, the giant crocodile *Deinosuchus*, and the mosasaur *Tylosaurus*. Another historical site in the Black Creek Formation was a marl pit southwest of Faison in Sampson

County, which produced the first *Hypsibema* bones, described in 1869 by Edward Cope.

Maryland (Severn Formation)

The Severn Formation has yielded the only Late Cretaceous dinosaur fossils in Maryland. In Prince Georges County east of Washington, D.C. (chapter 9), collectors have found evidence of indeterminate hadrosaurids and the ornithomimosaur *Coelosaurus,* as well as mosasaurs, plesiosaurs, crocodiles, and a soft-shelled turtle called *Trionyx.*

New Jersey/Delaware (Raritan Formation)

The Raritan Formation attains its best exposures in sand and clay pits along the Raritan River in Middlesex County, New Jersey, and southwest along the Delaware River. This formation contains both Lower and Upper Cretaceous sediments, and in New Jersey it is sometimes considered the upper part of the Potomac Group.

Only two dinosaur fossil finds are noted from the Raritan. The broken end of one of the foot bones of a large carnivorous dinosaur was recovered in the early 1900s in a sand pit at Roebling, Burlington County, New Jersey. (The stratigraphic source of this bone is somewhat in question: it may have come from the overlying Magothy Formation.)

The other discovery represents the only Cretaceous dinosaur footprints ever found on the East Coast. These tracks—probably made by a medium-sized theropod—were discovered in 1929–30 in the Hampton Cutter Clay Works Pit at Woodbridge, Middlesex County, New Jersey. The Raritan Formation has also yielded abundant plant fossils, lignite, and even some amber.

New Jersey/Delaware (Merchantville Formation)

This is the oldest greensand unit, consisting of a greasy dark blue or black glauconitic clay laid down in a shallow-water marine environment. The Merchantville Formation in New Jersey has yielded isolated remains of hadrosaurids from the Pennsylvania Clay Company pits at Matawan. Delaware's Chesapeake and Delaware Canal has produced a few fossils of a hadrosaurid and an ornithomimosaur from Merchantville sediments, as well as the marine turtle *Toxochelys* and the mosasaur *Tylosaurus.* Bones resembling the pterosaur *Pteranodon* were found in the Merchantville at the Chesapeake and Delaware Canal in 1981.

New Jersey (Woodbury Formation)

A marine formation of chocolate-colored clay, the Woodbury Formation outcrops in New Jersey from Raritan Bay to Gloucester County. It was the Woodbury Clay at Haddonfield, New Jersey, that yielded the world's first nearly complete dinosaur skeleton, described in 1858 as *Hadrosaurus foulkii*. Foot bones of *Ornithotarsus immanis* (now referred to *H. foulkii*) were found in 1869 at Keyport and in 1896 during excavation at Merchantville for a railroad underpass. A century earlier, a foot bone thought to be from a hadrosaurid dinosaur was found near Woodbury Creek and examined by members of the American Philosophical Society in Philadelphia (chapter 4).

New Jersey/Delaware (Marshalltown Formation)

The Marshalltown is one of the major greensand formations in New Jersey and Delaware, laid down approximately 75 Ma as the sea began another advance over the coastal plain. The prime New Jersey site was discovered in 1980 in the Crosswicks Creek area near Ellisdale in Monmouth County (chapter 9). The Ellisdale site has yielded teeth and bone fragments from the dinosaurs *Dryptosaurus* and *Hypsibema* as well as from indeterminate hadrosaurids, theropods, and the giant crocodile *Deinosuchus*. Theropod teeth have been collected from Marshalltown sediments at Mount Laurel and in the Monmouth Brook area. Marshalltown exposures at Freehold produced additional fossils of *Dryptosaurus* and *Hadrosaurus,* and bones of *Hadrosaurus* were found at Swedesboro. In Delaware, spoil from dredging of the Chesapeake and Delaware Canal contains Marshalltown sediments from which several hadrosaurid teeth and an ornithomimosaur toe bone have been collected.

New Jersey/Delaware (Wenonah/Mount Laurel Formations)

These marine sandstones are sometimes combined into one stratigraphic unit. The most productive dinosaur fossil site is along Big Brook, near Marlboro, New Jersey. Collectors at Big Brook have found bones of hadrosaurids, a nodosaurid ankylosaur, and the ornithomimosaur *Coelosaurus*. Other Big Brook fossils include fishes, marine turtles, and great marine lizards like *Mosasaurus maximus*. A site on Hop Brook near Holmdel, New Jersey, has produced remains of hadrosaurids and a theropod tooth, possibly *Dryptosaurus*. A hadrosaurid vertebra was discovered in Mount Laurel sediments at the Chesapeake and Delaware Canal.

New Jersey (Navesink Formation)

The Navesink Formation is a highly fossiliferous greensand formation in New Jersey that has produced more Late Cretaceous dinosaur material than any other formation on the East Coast. Another Upper Cretaceous unit known as the New Egypt Formation is commonly grouped with the Navesink. Many vertebrate fossils have been found in the Navesink "chocolate marl" sediments exposed at the Inversand Company's pit in Sewell, including a partial skeleton of what is now referred to as *"Hadrosaurus" minor,* plus primitive crocodiles, sea turtles, and mosasaurs. Other hadrosaurid remains have been found in New Jersey at Barnsboro, Marlboro, Atlantic Highlands, Swedesboro, Mullica Hill, and Big Brook. A single tooth found in 1865 at Mullica Hill was described as *Diplotomodon horrificus.* The *Laelaps* (now *Dryptosaurus*) skeleton described in 1866 came from the West Jersey Marl Company's pit in Barnsboro. Navesink sediments in Burlington County yielded the ornithomimosaur *Coelosaurus,* and evidence for a nodosaurid was found at Poricy Brook in Monmouth County.

New Jersey (Hornerstown Formation)

Buried within this greensand formation in New Jersey is the Cretaceous-Tertiary (K-T) boundary, the thin sediment layer that records the catastrophic extinction event at the end of the Cretaceous. (The *K* in K-T stands for *Kreide,* the German word for chalk, and refers to the great chalk accumulations formed in such Upper Cretaceous localities as the White Cliffs of Dover in England.) The lower part of the Hornerstown at Inversand recently yielded a single dinosaur fossil—a vertebra thought to belong to a hadrosaurid—and fossil birds, ammonites, and mosasaurs.

East Coast Dinosaurs of the Late Cretaceous

Information and insights about the Late Cretaceous dinosaurs of the East Coast come to us from the founding fathers of North American vertebrate paleontology—Joseph Leidy, Edward D. Cope, and Othniel C. Marsh—in the last century, and from modern workers including Donald Baird, Edwin H. Colbert, William B. Gallagher, John R. Horner, David C. Parris, James L. Knight, Edward M. Lauginiger, Robert K. Denton Jr., and others.

It is from their fieldwork and studies that we know sauropod dinosaurs were gone from our eastern realm, as in most parts of North America, by the

Late Cretaceous. Ornithopods had been greatly diminished in variety, although the duck-billed hadrosaurids were at their peak. Other groups of dinosaurs were also represented, including the ankylosaurs and ornithomimosaurs. And all these plant eaters lived in fear of the likes of the theropod *Dryptosaurus,* largest of the East Coast predators.

Ankylosauria: Last of the Armor

The armored ankylosaurs are the rarest dinosaur fossils from the Late Cretaceous of the East Coast. So far, all that is known is a tail vertebra collected from central New Jersey and a keeled scute (one of many bony plates formed in the skin) that appears to have been lost. Fortunately, a cast of it is still available for study.

As we learned in chapter 7, ankylosaurs are divided into two groups: those with tail clubs (ankylosaurids) and those without (nodosaurids). In addition to these anatomical differences, the two groups are distinguished by their scutes. For example, in ankylosaurids the base of the scutes is hollowed out and quite thin, but nodosaurids have scutes whose base tends to be flat or only slightly cupped. The New Jersey ankylosaur had scutes that are flat based, so it qualifies as a nodosaurid.

We must rely on better-known nodosaurids to provide insight into the behavior of the East Coast species. Our nodosaurids were probably about 5 to 6 meters (16 to 20 feet) long, with extensive body armor, huge spines over the shoulder, and no tail club. In the hunker-down mode of protection, these nodosaurids were well guarded against predators.

Ankylosaurs in general were low on brains and slow in speed (chapter 7), but they were probably more adept in the feeding arena. Like *Priconodon* from the Early Cretaceous of Maryland, they fed on shrubby plants, ground cover, and the leaves or fruits on low tree branches. After some brief chewing, this food was shipped off to the capacious gut for fermentation.

Ornithopoda: East Coast Duckbill Menagerie

The crowning glory of Ornithopoda were the hadrosaurids, or duck-billed dinosaurs, the most abundant of dinosaur finds in the Upper Cretaceous rocks of the East Coast. North America's first relatively complete dinosaur was *Hadrosaurus foulkii,* discovered at Haddonfield, New Jersey, and described in 1858 (chapter 4).

We know quite a bit about hadrosaurids, thanks largely to discoveries in the North American West of well-preserved, fully articulated skeletons.

Western sites have also produced numerous nests of hadrosaurid eggs (some with embryos) and a wide range of remains that includes hatchlings, juveniles, "teenagers," and adults. Numerous hadrosaurid footprints and trackways have been found in other parts of the world.

All these duckbills were large—10 meters (33 feet) long—and had heads with a broad beak and jaws with hundreds of teeth. In some hadrosaurids the top of the head was flat and in others it was topped by a bony or hollow crest. The large body was supported by muscular hind legs, although the smaller forelegs may have been used when the animal was resting or walking slowly. Otherwise hadrosaurids were bipedal. In keeping with this kind of locomotion, the long tail projected straight back from the rear of the body to counterbalance the front.

What about our East Coast hadrosaurids? The most famous is *Hadrosaurus foulkii* ("Foulke's bulky reptile"), named by Leidy for William Parker Foulke (chapter 4). As the namesake for the ornithopod group known as Hadrosauridae, *Hadrosaurus foulkii* is important because of its role in the early history of dinosaur paleontology, and even more because of its value to modern studies of hadrosaurids.

Hadrosaurus foulkii is still the most completely known of any East Coast dinosaur (fig. 8.1). In addition to the Haddonfield specimen, there are remains of this species—including a partial hind leg, a metatarsal, and lower jawbones—from several other sites in New Jersey. Some of this material was originally called *Ornithotarsus immanis* ("enormous bird tarsus") and *Hadrosaurus cavatus* (*cavatus* means "hollow") by Cope.

Hadrosaurus foulkii is not the only hadrosaurid from the Late Cretaceous of the East Coast. Others include *Hadrosaurus minor* (*minor* means "little"), which consists of trunk and tail vertebrae from Barnsboro, New Jersey, described by Marsh in 1870. This material is now thought to be too fragmentary to deserve its own genus and species names. To make matters more confusing, a nearly complete rear half of a skeleton from Sewell, New Jersey, was given the same name by Edwin H. Colbert in 1948. Many paleontologists think this dinosaur should be renamed, so we will simply call it "*Hadrosaurus*" *minor*, with quotation marks to indicate the uncertain assignment. Other isolated material—a femur, a tibia, a vertebra—of these latest Cretaceous hadrosaurids has also been tentatively assigned to "*Hadrosaurus*" *minor*.

Two other hadrosaurids deserve notice. The first, one of the largest of North American hadrosaurids, is the 12 meter (39 foot) long *Hypsibema crassicauda* ("high step, thick tail"), described by Cope in 1869. *Hypsibema* has been a puzzle since its discovery in North Carolina in the mid-1800s. The original agglomeration of water-worn bones from several marl pits was thought to represent a single large hadrosaurid. A 1979 study by Donald Baird and John Horner (chapter 9) sorted out this material, allotting some of it to theropods, some to hadrosaurids, and the rest—the tail vertebrae that form the basis for the species name of *Hypsibema*—to a sauropod. However, recent views are tending back toward the original assignment of the remains to a true hadrosaurid. Another North Carolina find, a lower-jaw tooth, has affinities with *Lophorhothon* from Alabama, a small hadrosaurid 5 meters (16 feet) long.

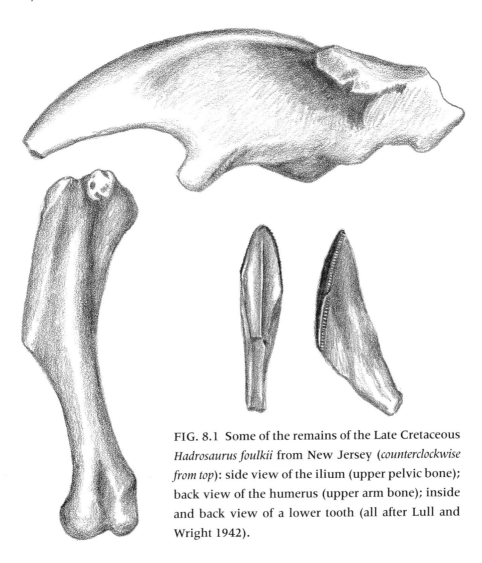

FIG. 8.1 Some of the remains of the Late Cretaceous *Hadrosaurus foulkii* from New Jersey (*counterclockwise from top*): side view of the ilium (upper pelvic bone); back view of the humerus (upper arm bone); inside and back view of a lower tooth (all after Lull and Wright 1942).

FIG. 8.2 The hadrosaurid *Gryposaurus,* from the Late Cretaceous of Alberta, Canada.

These East Coast hadrosaurids, like their duck-billed relatives elsewhere in the world, were fully terrestrial, not amphibious as originally thought. That early mistake was understandable for a number of reasons. Most of the early discoveries were in rocks of marine origin, and the remains revealed that the tail was long and deep, the hand looked webbed, and the jaws appeared too weak to handle anything but soft aquatic vegetation.

It turns out that the long, deep tail—once thought to be a sort of rudder for the swimming dinosaur—was constructed to be held nearly horizontal, counterbalancing the front of the body (fig. 8.2). This horizontal body posture and bipedal stance characterize *Hadrosaurus foulkii,* "*Hadrosaurus*" *minor,* and all other hadrosaurids. The webbed hand, upon close scrutiny, isn't a web at all. Instead, the three hoofed fingers were joined together in a thickened pad. Much like modern camels, the padded hand functioned solely to support the body while the animal was standing on all fours.

How fast could these dinosaurs have traveled in their more common bipedal stance? *Hadrosaurus foulkii* and "*Hadrosaurus*" *minor* may have cruised the countryside at 4 to 5 meters a second (9 to 11 miles an hour). They appear to have been capable of 13 meters a second (29 miles an hour) over short distances, especially when chased by a large predator!

Hadrosaurids, like all ornithopods, probably spent a good deal of time feeding. As we learned with ?*Tenontosaurus* from the Lower Cretaceous Arundel Clay (chapter 7), most ornithopods had no front teeth. They had broad snouts and could grind up plant food by allowing their upper jaws to rotate slightly.

Dinosaurs like *Hadrosaurus foulkii* probably ate great quantities of highly fibrous foliage from low conifers or angiosperm shrubs and trees. It is likely that all hadrosaurids had large guts to ferment and digest low-quality vegetation.

Were these fast movers and good eaters smart too? Hadrosaurids in general appear to have had larger braincases than most other dinosaurs. Large brain size may relate to the complex behavior ascribed to hadrosaurids. For example, from the time these dinosaurs were first discovered, their many and varied cranial ornaments have fascinated paleontologists. These wild headdresses were once thought to relate to aquatic habits or to the animals' sense of smell. But research by James A. Hopson of the University of Chicago favors combat and social display as the most compelling explanation for the evolution of these unusual structures.

Hadrosaurus foulkii, which may have had an accentuated arch snout like *Gryposaurus,* could have used its head for broadside or head-on pushing contests. In addition, an inflatable flap of skin may have covered the nostril region; this flap could be inflated and used for visual display, as well as to make a noise like bagpipes.

The ultimate hadrosaurid displays belonged to the lambeosaurines. The hollow crests atop the heads of *Corythosaurus* and *Parasaurolophus* were so distinctive that they must have provided for instant recognition, either visually or by the low honking sounds that resonated within them.

This kind of display and combat between members of the same species makes even more sense when considered with other aspects of their behavior. A number of sites in the American and Canadian West are great jumbles of bone dominated by single hadrosaurid species, suggesting that these animals were very common and perhaps formed herds of youngsters and adults. This herding behavior may also apply to the hadrosaurids of the East Coast, the commonest dinosaurs from the Late Cretaceous.

Other hadrosaurid research, principally by John R. Horner, indicates that some forms—*Maiasaura* and *Hypacrosaurus,* among others—probably nested in huge colonies. Upon hatching, hadrosaurid babies were nest bound for as long as eight or nine months. One or both parents brought food to the nest and protected the youngsters from predators.

So hadrosaurids had plenty of opportunity to interact: in herds, as breeding pairs, and as families. On this basis alone, we shouldn't be surprised to find the great variety of skulls and crests that feature in the arena of social behavior among these complex animals.

FIG. 8.3 (*left*) Tibia (shinbone) of the ornithomimosaur *Coelosaurus antiquus,* from the Late Cretaceous of New Jersey and Delaware.

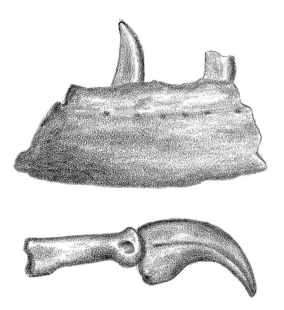

FIG. 8.4 Some of the remains of the Late Cretaceous *Dryptosaurus aquilunguis* from New Jersey: (*top*) outside view of the lower jaw, with two teeth; (*bottom*) the last two bones, including the giant claw, of the first finger.

Theropoda: "Old Hollow" and "Tearing Eagle Claw"

The theropods from the Late Cretaceous of the East Coast are better known than those from any earlier period. In addition to footprints, there is skeletal material of a small ornithomimosaur known as *Coelosaurus antiquus* ("old hollow reptile") from central New Jersey and northern Delaware (fig. 8.3). Then there's the larger and fiercer *Dryptosaurus aquilunguis* ("tearing reptile, eagle claw"), known from a partial skeleton and isolated bones and teeth (fig. 8.4), also from central New Jersey. Finally, an unusually broad and straight tooth given the name *Diplotomodon horrificus* ("dreadful double-cutting tooth") by Leidy in 1865 provides a fleeting glimpse of another Late Cretaceous theropod from the East Coast.

The names of all three of these theropods have had a somewhat checkered history. The name *Coelosaurus,* originally assigned in 1865 by Leidy, had already been used by Sir Richard Owen in 1854 for another fossil reptile. Baird and Horner rediscovered Owen's prior use and suggested that all East Coast ornithomimosaur material from the Late Cretaceous be called *Ornithomimus.* Unfortunately, they never made the case that the East Coast *Coelosaurus* was the same thing as *Ornithomimus,* otherwise known only from the Western Interior of North America. For this reason and because few if any subsequent paleontologists have ever used Owen's name *Coelosaurus,* we have chosen not to refer to the Late Cretaceous ornithomimosaur as *Ornithomimus antiquus,* instead sticking with Leidy's original name, *Coelosaurus antiquus.*

Diplotomodon was originally christened *Tomodon* when Leidy first described it in 1865 as a plesiosaur, a long-necked oceangoing reptile. But this name had already been used, so Leidy gave the same tooth the name *Diplotomodon* in 1868, thinking by then that it was a fish. This tooth later was attributed to a mosasaur, quite common in the marine beds of the region. It wasn't until 1952 that *Diplotomodon* was "transformed" into a theropod dinosaur by Samuel P. Welles of the University of California at Berkeley. To make matters worse, the tooth is now lost, making this poorly understood theropod completely unexaminable.

Finally, turning to *Dryptosaurus,* this theropod was first named *Laelaps* (for the mythical hunting dog Laelaps, or "storm wind") when it was described by Cope in 1866. Because this name had already been used for an insect, Marsh—Cope's archrival (chapter 4)—replaced it with *Dryptosaurus* in 1877, and it has been *Dryptosaurus* ever since.

All these East Coast theropods represent fairly specialized predators. For much of the time since the discovery of *Dryptosaurus* in 1866, this largest of East Coast theropods was thought to be one of the giant meat-eating dinosaurs known as carnosaurs. But with the reinvigoration of research on theropod evolution, Carnosauria has disappeared as a group with a unique history. So *Dryptosaurus* can't be a carnosaur.

In 1990 Robert K. Denton (chapter 9) first suggested that *Dryptosaurus* may have been an exceptionally large member of the group of theropods called Coelurosauria. Such evolutionary affinities are borne out by a special structure of the ankle and foot, features shared by the ornithomimosaurs, *Dryptosaurus, Troodon,* and *Tyrannosaurus.* Based on those features, Thomas R. Holtz Jr. of the U.S. Geological Survey has called this great group of theropods Arctometatarsalia ("compressed metatarsus"). So *Dryptosaurus*

FIG. 8.5 Reconstruction of *Dryptosaurus*.

goes from being an East Coast enigma to being an East Coast arcto-metatarsalian.

Work on the evolutionary position of *Dryptosaurus* is bound to continue, but we can be more certain about its lifestyle during the Late Cretaceous. First of all, *Dryptosaurus* was a biped 4 to 5 meters (12 to 15 feet) long (fig. 8.5). Like all other theropods, the front of the body was balanced directly over the hips by a long, well-muscled tail, and the backbone was probably held near-ly horizontal. The hind legs were strong, and the feet were tipped with ex-tremely sharp claws.

From limb proportions alone, it is clear that *Dryptosaurus* was a swift preda-tor, perhaps reaching speeds up to 5 meters a second (11 miles an hour). Prey of any sort would have had a hard time escaping its powerful hind legs and sharply clawed feet. The latter were especially good for disemboweling prey, as were its strong arms and its hands with huge, scimitarlike recurved claws. Death came swiftly at the hands and feet of this large theropod.

But as with nearly all theropods, it was the jaws and teeth that delivered the coup de grâce. The pointed and sharply serrated teeth of *Dryptosaurus* were always ready to slice and tear through flesh, muscle, and sinew. These weapons were augmented by small pockets at the base of each serration that may have harbored bacteria to promote infection, shock, and ultimate death in victims.

Life must have been considerably less bloodthirsty for *Coelosaurus*. As in other ornithomimosaurs (chapter 7), the swiftest of all dinosaurs, speed provided the sole escape from predators like *Dryptosaurus*. Fruits, grubs, insects, and an occasional small lizard or mammal provided *Coelosaurus* with a pleasant life between the onslaughts of terror.

End of the Late Cretaceous

These were the dinosaurs that occupied the East Coast stage as the end of the Cretaceous approached: *Hadrosaurus foulkii,* "*Hadrosaurus*" *minor, Diplotomodon horrificus, Coelosaurus antiquus, Dryptosaurus aquilunguis,* and one or more nodosaurid ankylosaurs. Sharing the scene were a great many freshwater and saltwater crocodilians, a few tiny mammals, a giant pterosaur, lizards and turtles, and—in keeping with the marine rocks of the region—lots of mosasaurs, fishes, and oceangoing invertebrates, including clams, snails, and ammonites.

Some 65 Ma, here on the East Coast and apparently around the globe, life on Earth changed drastically. Nearly all of the dinosaur groups and many others suffered great extinction. How could such devastation have happened? The cause of the great Cretaceous-Tertiary mass extinction and other dinosaur mysteries still puzzling paleontologists will be explored in chapter 10.

FIG. 9.1 Dinosaur National Monument in Jensen, Utah. Reliefing work shown in progress on the quarry face has since been completed. (National Park Service photograph)

9

THE MODERN SEEKERS

The Big Sleep

There's no one date that marks the end of the golden age of dinosaur hunting on the East Coast. Collecting by Arthur Bibbins in Maryland's Arundel Clay from 1894 to 1896 was one of the last fossil-hunting expeditions in eastern North America supported by a college or museum until the mid-1900s. The 1910 discovery of *Podokesaurus* by Mignon Talbot at South Hadley, Massachusetts, represents the only nearly complete dinosaur skeleton found on the East Coast in the entire twentieth century (chapter 4).

The decline of collecting in the East resulted from a number of factors, both economic and scientific. As we mentioned in previous chapters, the quarrying industries that opened such productive exposures in the fossil-bearing sediments gradually shut down. Brownstone in the Connecticut Valley, iron ore in Maryland, greensand in New Jersey—all suffered major drop-offs in demand after peaking in the late 1800s.

In addition, the focus of dinosaur paleontology had begun shifting in the 1870s to the newly discovered fossil fields of the American and Canadian West. Dinosaur fossil pickings in the East seemed meager and hard won compared with the fabulous skeletons of new species that were literally falling from the well-exposed rocks in Utah, Montana, and Alberta, Canada (fig. 9.1).

Eastern museums and universities put their resources into collecting expeditions likely to produce the most bone for the buck.

But not even the lure of the great western sites was enough to forestall a continentwide slowdown in exploration caused by the Depression and the two world wars. Dinosaur paleontology languished, and the pace of discovery slowed.

On the East Coast, the hunt was kept alive during the first six or seven decades of the twentieth century by a handful of dedicated researchers and collectors. Many of them also pursued important work elsewhere in North America and abroad, but they appreciated the unique resource in their own backyards and continued to probe the great fossil secrets of the East.

The True Believers

Richard Swann Lull carried on the Connecticut Valley footprint research begun by Edward Hitchcock in the 1830s. Lull earned his doctorate from Columbia University in 1903 and enjoyed a long career at Yale University, where he was a popular professor of paleontology and director of the Peabody Museum of Natural History. He began publishing on Connecticut Valley dinosaurs in 1904, but it was his classic *Triassic Life of the Connecticut Valley* (1915) that established his authority on the subject. Among his other East Coast publications was a 1911 paper on the Early Cretaceous dinosaurs of the Arundel Clay in Maryland and a 1953 revision of *Triassic Life.*

Charles Whitney Gilmore was a dinosaur specialist and curator at the U.S. National Museum of Natural History (Smithsonian Institution) in Washington, D.C. Gilmore collected widely and named nine new dinosaur species. On the East Coast, he made major revisions in 1921 to the Arundel dinosaurs of Maryland, studied earlier by Marsh and Lull. In 1942 he helped recover a dinosaur almost on his doorstep: a large *Astrodon* thighbone found during excavation for a water filtration plant in Washington (chapter 4).

Friedrich von Huene was primarily known for his pioneering research into the Triassic life of Europe and South America. But the German paleontologist also pursued East Coast vertebrate paleontology in the early 1900s. In 1913 von Huene named the phytosaur *Rutiodon manhattanensis* based on a skeleton found near the Palisades at Fort Lee, New Jersey. In 1922 he named the type species of *Stegomosuchus longipes,* a primitive crocodile from Longmeadow, Massachusetts. In 1932 he revised the Connecticut Valley prosauropod skeleton *Anchisaurus colurus* as *Yaleosaurus* (now *Anchisaurus polyzelus*).

FIG. 9.2 Edwin Colbert of the Museum of Northern Arizona is one of the great popularizers of dinosaurs and a researcher on East Coast dinosaurs. (Courtesy of E. H. Colbert)

The number of new dinosaur species described by von Huene stands at twenty-seven, including fourteen named all in one year: 1932.

Wilhelm Bock, an engineer by training, was an accomplished paleontologist associated with the Academy of Natural Sciences of Philadelphia from the 1940s to the early 1960s. He collected and described Triassic fishes, plants, and reptiles from such famous sites as the railroad cut in Gwynedd, Pennsylvania, the Smith Clark Quarry in Milford, New Jersey, and the Squirrel Hill Quarry near Schwenksville, Pennsylvania. His most important work, "Triassic Reptilian Tracks and Trends of Locomotive Evolution, with Remarks on Correlation," was published in 1952. Bock named fifty-eight plant species, nine invertebrate species, and twenty-five vertebrate species, including many dinosaur footprints.

Edwin H. Colbert, now affiliated with the Museum of Northern Arizona in Flagstaff, has been a tireless popularizer of dinosaur paleontology for more than fifty years, as well as a researcher on both eastern and western dino-

FIG. 9.3 Donald Baird chisels a dinosaur tooth from a sandstone cliff near Noel, Hants County, Nova Scotia, in 1967. (Courtesy of Donald Baird)

FIG. 9.4 Donald Baird, the leading authority on the footprints and bones of East Coast dinosaurs, poses in 1982 with the original plaster *Hadrosaurus foulkii* skull modeled by Waterhouse Hawkins. (Courtesy of Richard C. Ryder)

saurs (fig. 9.2). After earning his doctorate from Columbia University in 1935, Colbert began a careerlong association with his alma mater, teaching zoology and vertebrate paleontology. He also served as curator of fossil reptiles and amphibians at the American Museum of Natural History in New York from 1943 to 1970. Colbert has published many important papers and books on East Coast paleontology, including the popular *Fossils of the Connecticut Valley: The Age of Dinosaurs Begins* in 1963. Major eastern discoveries described by Colbert began in 1946 with a skeleton of the procolophonid *Hypsognathus fenneri* from Passaic, New Jersey. Others include *"Hadrosaurus" minor* from Sewell, New Jersey, in 1948; the phytosaur *Rutiodon carolinensis* from North Bergen, New Jersey, in 1965; and the gliding reptile *Icarosaurus siefkeri*, also from North Bergen, in 1966.

Donald Baird is the leading authority on the footprints and skeletal remains of dinosaurs on the East Coast (figs. 9.3, 9.4). Baird earned his Ph.D. in biology in 1955 from Harvard University and in 1957 joined the staff of the Museum of Geology at Princeton University. He made his long and productive career at the museum, retiring as director in 1988. Baird's research on eastern fossil reptiles and amphibians began in 1954 with his publication on the footprint *Chirotherium lulli*. He has collected and described a host of dinosaur fossils from the Carolinas to Nova Scotia.

Baird and Colbert have helped inspire a generation of younger researchers to push ahead in understanding the creatures of the Mesozoic, learning what they can from footprints, teeth, and skeletons. The rebirth of dinosaur paleontology on the East Coast was made possible by all the true believers who kept the torches lit during the dark middle decades of this century.

New Golden Age

The spark widely credited with rekindling the excitement of dinosaur hunting after the Depression and World War II was the discovery in Montana in 1964 of a startling new theropod dinosaur. John H. Ostrom of Yale University and his team found the skeleton of a fierce human-sized carnivore armed with bladelike teeth and razor-sharp claws. In his description of *Deinonychus* in 1969, Ostrom (fig. 9.5) concluded that this dinosaur was a swift and agile predator, highly successful at the hunt, probably warm-blooded and closely related to the birds (chapter 10).

This contradicted the traditional view of dinosaurs as cold-blooded, sluggish, and stupid. Another major blow to that stereotype came in 1978 when

FIG. 9.5 John Ostrom pioneered the modern view of dinosaurs with his discovery in 1964 of the agile, intelligent *Deinonychus* from Montana. (Courtesy of T. Charles Erickson, Yale University, Office of Public Information)

FIG. 9.6 Jack Horner, curator of paleontology at the Museum of the Rockies, found fossilized dinosaur eggs and nests in Montana in 1978. (Museum of the Rockies/Bruce Selyem)

John R. Horner and Robert Makela found fossil dinosaur eggs and nests in Montana that suggested maternal behavior in a hadrosaurid dinosaur they named *Maiasaura*. Horner (fig. 9.6), now curator of paleontology at the Museum of the Rockies in Bozeman, Montana, was then a research assistant to Baird at Princeton University.

Ostrom had studied under Colbert at Columbia University, where he earned a doctorate in geology in 1960. He began his careerlong association with Yale University a year later. From 1971 until his retirement at the end of 1993, Ostrom was professor of geology and geophysics at the university and curator of vertebrate paleontology at the Peabody Museum of Natural History. In addition to *Deinonychus,* Ostrom named the theropod *Microvenator,* the nodosaurid ankylosaur *Sauropelta,* and the ornithopod *Tenontosaurus,* all in 1970.

With the intellectual ferment surrounding the *Deinonychus* discovery came new interest in the dinosaurs of the East Coast. Discoveries of new fossil sites and renewed attention to old sites reestablished the East as an important and exciting chapter in the dinosaur story.

Connecticut Valley

The great Triassic-Jurassic rift valley in Massachusetts and Connecticut that produced the first dinosaur fossils on the East Coast still yields footprints and bones for those who look. As Richard D. Little wrote in *Dinosaurs, Dunes, and Drifting Continents: The Geohistory of the Connecticut Valley* (1984): "Every split of a sedimentary rock exposes a never-before-seen page in the ancient history of our Valley, perhaps revealing secrets of former inhabitants or their environment."

Dinosaur State Park

On August 24, 1966, workmen made a remarkable discovery while excavating a site in Rocky Hill, Connecticut, near Hartford, for a new State Highway Department testing laboratory. About 4 meters (12 feet) below the surface was a rock layer that had held a secret treasure for 185 million years: hundreds of three-toed footprints ranging from 25 to 40 centimeters (10 to 16 inches) long (fig. 9.7). Work was halted at the site, and Ostrom and other specialists were called in to assess the importance of the discovery.

They found nearly 1,500 well-preserved tracks and dozens of trackways, most made by a theropod dinosaur approximately 2.5 meters (8 feet) tall and

FIG. 9.7 Dinosaur footprint site at Rocky Hill, Connecticut, when it was first uncovered in 1966. (Dinosaur State Park)

FIG. 9.8 Closeup of *Eubrontes* print (approximately 30 centimeters [12 inches] long) at Dinosaur State Park, Rocky Hill, Connecticut. (Dinosaur State Park)

FIG. 9.9 Richard Krueger, environmental education coordinator at Dinosaur State Park.

6 meters (20 feet) long. The most abundant track (fig. 9.8) is known as *Eubrontes*, made by a theropod similar to *Dilophosaurus*. Three-toed *Grallator* tracks at the site are attributed to a smaller theropod like *Coelophysis*, and five-toed *Batrachopus* prints were made by a primitive crocodile-like reptile. Numerous mud cracks, raindrop impressions, and ripple marks suggest that the Rocky Hill site was a muddy lakeshore in Early Jurassic times.

Ostrom published the discovery in 1967. Soon after, the original exposure was reburied to help preserve it. A later excavation at the site revealed an additional five hundred tracks. In 1980 Walter P. Coombs Jr., a paleontologist at Western New England College in Springfield, Massachusetts, derived important new evidence for swimming ability in theropod dinosaurs from a curious trackway on the new exposure. The series of imprints suggests a dinosaur paddling along in shallow water, kicking the muddy bottom with the tips of its toes (chapter 6).

Since 1968 the Rocky Hill site has been preserved for posterity as Dinosaur State Park, and *Eubrontes* has since been honored as Connecticut's state fossil. Since 1970, Richard L. Krueger (fig. 9.9) has served as on-site environmental education coordinator. A permanent geodesic dome was erected over the exposed

FIG. 9.10 Exterior of Dinosaur State Park dome, built in 1977.

prints in 1977 (fig. 9.10) and houses educational displays and research facilities. A popular attraction at Dinosaur State Park is the "make your own dinosaur footprint" feature. Visitors who bring their own plaster casting supplies can make a mold of a *Eubrontes* track on a rock slab outside the dome. The well-worn three-toed footprint may be the most familiar dinosaur fossil in North America.

Manchester Bridge

Another Connecticut Valley discovery in the late 1960s finally laid to rest some unfinished dinosaur business from the previous century. As related in chapter 4, a prosauropod dinosaur skeleton found in 1884 at the Wolcott Quarry in Manchester, Connecticut, was cut in two during quarrying operations. A stone block containing the hind limbs and pelvis of the dinosaur was recovered, but blocks that may have held the skull and forelimbs were carted off and subsequently built into a bridge abutment in South Manchester.

The story might have ended there, except for the efforts eighty-five years later of Yale's John Ostrom. "Ever since 1884," Ostrom wrote, "curators of vertebrate paleontology at the Peabody Museum have kept a hopeful eye on the brownstone bridge in Manchester, waiting for its demolition." The long wait came to an end in August 1969, when the 40 foot long bridge (fig. 9.11) over Hop Brook at Bridge Street in South Manchester was finally demolished (fig. 9.12) to make way for a new span. Ostrom and his team from the Peabody were on the scene to search for the missing blocks. "Some 400 sand-

FIG. 9.11 Bridge in Manchester, Connecticut, built in 1884 with stone blocks containing part of a dinosaur skeleton. (Courtesy of the Peabody Museum of Natural History, Yale University)

FIG. 9.12 The demolition of the Manchester bridge in 1969, when a missing part of *Ammosaurus* skeleton was recovered by John Ostrom. (Courtesy of the Peabody Museum of Natural History, Yale University)

stone blocks were examined during the two-day project, and two were found that contained fossil bone," Ostrom later recalled. One of those 500-pound blocks held the missing half of the right femur of *Ammosaurus major,* which has since been reunited with the original skeleton at the Peabody Museum.

Mt. Tom

The Mt. Tom area near Holyoke, Massachusetts, has long been known for its dinosaur and other reptile tracks described by Edward Hitchcock in the early nineteenth century and later by Richard Swann Lull. But it wasn't until 1970 that a wealth of dinosaur trackways at Mt. Tom were identified and systematically mapped, again by John Ostrom of Yale University.

Ostrom described the site in 1972 as a short distance north of Holyoke and east of the Mt. Tom Mountain Park recreation area, along the west shore of the Connecticut River. The track-bearing surface is approximately 40 meters (150 feet) long by 20 meters (60 feet) wide, populated by at least 134 three-toed footprints ranging in length from 9 centimeters (3.5 inches) to 35 centimeters (13.8 inches). The theropod dinosaur trackmakers are the same cast of characters as from Rocky Hill and many other footprint sites in the Connecticut Valley during the Early Jurassic: *Grallator* and *Eubrontes.*

The Mt. Tom site holds a special place in East Coast dinosaur paleontology because the trackways may provide rare evidence for flocking or herding behavior in theropods. Ostrom identified twenty-eight separate trackways, and approximately 70 percent are oriented in nearly parallel courses, suggesting that these dinosaurs were traveling together in a large group across a broad mudflat.

New Jersey

New Jersey, source of the world's first nearly complete dinosaur skeleton, remains a leader in dinosaur paleontology on the East Coast. Thanks to such active groups as the Delaware Valley Paleontological Society and the Monmouth Amateur Paleontologists Society, the New Jersey area has been combed for fossils more thoroughly than any other region in eastern North America. Its large and diverse inventory of dinosaur fossils is a major resource for researchers, who are continually updating dinosaur science in the light of new discoveries. Today, dedicated collectors continue the state's long tradition of important dinosaur discoveries at a number of sites in New Jersey.

FIG. 9.13 The greensand marl pit at Sewell, New Jersey, operated by the Inversand Company, shown in 1993.

Sewell

The only surviving commercial greensand pit in New Jersey has produced more important Cretaceous fossils than any other site in the state. The treasures include duck-billed dinosaurs, primitive crocodiles, mosasaurs, sea turtles, and birds. Situated in an outcrop of nearly pure glauconite, the marl pit at Sewell was opened in the 1860s and is now operated by the Inversand Company to produce greensand for water conditioning (chapter 3).

The Sewell site (fig. 9.13) exposes sediments formed during the Late Cretaceous and early Tertiary, bridging the great extinction event at the end of the Cretaceous that killed off the dinosaurs (chapter 10). Greensands were deposited on the floor of an ancient ocean that once covered the area to a depth of up to 90 meters (300 feet).

Although many historic dinosaur fossil finds were made within 16 kilometers (10 miles) of the quarry—at such sites as Swedesboro, Mullica Hill, and Barnsboro—the first dinosaur discovery at Sewell came in the fall of 1947. Workers at the pit noticed the ancient bones and alerted Horace G. Richards, then curator of geology and paleontology at the Academy of Natural Sciences of Philadelphia. Richards and Edwin Colbert of the American

Museum of Natural History directed removal of the partial skeleton, which Colbert described in 1948 as *Hadrosaurus minor.* This is the same species we refer to as *"Hadrosaurus" minor* in chapter 8.

In 1957 a 24 inch long hadrosaurid tibia, dated at 70 million years old, was identified from the Sewell pit by Glenn L. Jepsen of Princeton University. In addition, two nearly complete skulls of *Mosasaurus maximus,* a 40 foot long seagoing lizard, were uncovered in 1961. Mosasaur remains are among the most common Cretaceous reptile fossils in New Jersey.

Recent collecting at the pit has been led by staff from the New Jersey State Museum in Trenton. David C. Parris, curator of natural history, and William B. Gallagher, registrar of natural history, have both conducted research at Inversand. Parris (fig. 9.14), a paleontologist and geologist known primarily for his work on Cretaceous birds, has also collected and studied many marine forms, including turtles, crocodiles, and mosasaurs.

Gallagher (fig. 9.15) earned his doctorate in geology at the University of Pennsylvania and has collected widely in the United States and abroad. He cofounded the Delaware Valley Paleontological Society in 1978, served as its first president, and was editor of the society's journal, *The Mosasaur,* one of the leading publication outlets in eastern North America for dinosaur research. Gallagher's 1990 state museum bulletin, *Dinosaurs, Creatures of Time,* stands as the best popular look at the dinosaurs of New Jersey.

Dinosaur fossils from the Inversand pit are rare discoveries. Since the 1930s the museum has had an arrangement with the quarry management to save the most valuable fossils found by workers. In 1980 Parris collected a hadrosaurid lower front leg bone from the pit. Gallagher found a hadrosaurid vertebra at Inversand in 1988, and in 1989 he unearthed a fragmented leg bone that may represent the largest carnivorous dinosaur bone found in New Jersey since Edward Cope described *Laelaps* (now *Dryptosaurus*) in 1866. A more recent find by Gallagher came in 1991, when he discovered another vertebra thought to belong to a hadrosaurid.

Riker Hill Quarry

In 1968 a fourteen year old fossil sleuth named Paul E. Olsen from Livingston, New Jersey, was exploring the Riker Hill Quarry in nearby Roseland with a friend, Tony Lessa. Their diligent efforts were rewarded with the discovery of Early Jurassic dinosaur footprints (fig. 9.16). Olsen—an avid fossil hunter since he was eleven years old—and Lessa led a crusade to have the fossil site preserved. In 1970 Walter Kidde and Company, Inc., owner of

FIG. 9.14 Dave Parris of the New Jersey State Museum at work in Brule County, South Dakota, in 1990. (Courtesy of William B. Gallagher)

FIG. 9.15 Bill Gallagher of the New Jersey State Museum at the Inversand Company marl pit in Sewell, New Jersey, in 1993.

FIG. 9.16 Teenagers Paul Olsen (*right*) and Tony Lessa pose in 1969 with a rock slab imprinted with *Anomoepus* dinosaur tracks, collected at the Riker Hill Quarry in Roseland, New Jersey. (Courtesy of Robert Salkin)

FIG. 9.17 Paul Olsen of Columbia University, a specialist on the Newark Supergroup, in 1990 next to core samples from drilling project in the Newark Basin. (Courtesy of Bruce Cornet)

the quarry, donated nineteen acres to Essex County, New Jersey, and the site is now administered by the county parks commission. Although an exhibit of Riker Hill dinosaur tracks was later stolen from the site, there remains a productive footprint exposure at the quarry that may be visited by permission.

The Early Jurassic dinosaurs at Riker Hill left at least three kinds of tracks: *Anomoepus, Grallator,* and *Eubrontes.* In addition, there are tiny tracks that may have been made by baby dinosaurs and a series of five-toed footprints that could represent the oldest record of primitive mammals in the Northern Hemisphere. The Riker Hill Quarry has also yielded fossils of plants, fishes, insects, and dinosaur coprolites.

Olsen, while still an honors undergraduate in geology at Yale University in 1975, wrote a paper on the Riker Hill discovery. He earned a doctorate in biology from Yale in 1984 and taught for six years at Columbia University in New York. In 1991 he joined Columbia's Lamont Doherty Earth Observatory in Palisades, New York, where he is now an associate professor of geology (fig. 9.17).

Olsen is best known as a specialist on the Newark Supergroup, the series of exposed Triassic-Jurassic basins extending along the East Coast from South Carolina to Nova Scotia (chapter 3). Since 1990 he has been conducting an extensive core-drilling project in these basins with Lamont Doherty colleague Dennis V. Kent. Building on the earlier work of such researchers as Franklyn B. Van Houten, professor emeritus of geology at Princeton University, Olsen has helped make the Newark Supergroup one of the most carefully studied and correlated geologic record in North America.

Ellisdale

One of the best examples of the unique nature of East Coast dinosaur paleontology is the Late Cretaceous site discovered in 1980 at Crosswicks Creek near Ellisdale, New Jersey (fig. 9.18). The Crosswicks Creek Basin has long been known for its exposures of the Marshalltown Formation, which has yielded Late Cretaceous dinosaur fossils from several areas in New Jersey and from the Chesapeake and Delaware Canal in Delaware.

In the summer of 1980 Robert K. Denton Jr. and Robert C. O'Neill (fig. 9.19) went prospecting for fossils along Crosswicks Creek and its tributaries in Ellisdale, hoping to strike it rich in the Marshalltown sediments. They had been inspired by the collection of Marshalltown fossils discovered in Delaware by Ralph O. Johnson Jr. of the Monmouth Amateur Paleontolo-

FIG. 9.18 The Upper Cretaceous Ellisdale microfossil site in New Jersey in 1993.

FIG. 9.19 Bob Denton (*right*) and Bob O'Neill, discoverers of the Ellisdale site.

gists Society. Denton is a physical chemist in New Brunswick, New Jersey, with paleontological field experience in Wyoming and Montana. O'Neill is a driver for the U.S. Postal Service and a fossil enthusiast who has been collecting with Denton since 1978.

Along a tributary to the creek, they found lignitized wood and pieces of fossil bone, eroded out of the steep banks. Over the next four years, after each heavy rain, they visited on weekends and collected dozens of vertebrate specimens. That inventory alone made Ellisdale one of the most important new Late Cretaceous fossil sites on the East Coast. But Denton and O'Neill were in for a pleasant surprise. In March 1984 a torrential storm dumped nearly 20 centimeters (8 inches) of rain on the central New Jersey area, and the little tributary at Ellisdale roared to life. When the "two Bobs" later checked their site, they found the streambed littered with bone fragments. Within a month, they had collected three hundred specimens of dinosaurs, turtles, crocodiles, bony fishes, lizards, and mammals.

Denton coordinated his fieldwork at the site with Dave Parris and with research associate Barbara S. Grandstaff of the New Jersey State Museum. The first report on Ellisdale was made in 1985, and collecting began in earnest in 1986. To date, more than 10,000 specimens have been recovered. Most are of terrestrial or freshwater origin, making the site very similar in character to the Phoebus Landing locality in North Carolina.

Ellisdale is known as a "microfossil" site. There are no skeletons or whole bones to be found. The specimens at Ellisdale were broken into small pieces as they were transported to their final resting place 75 Ma. Denton pictures the site in Late Cretaceous time as a channel or lagoon shoreward from a barrier island system, with sediments deposited by powerful storm surges from the sea. Today the Marshalltown Formation is exposed just above the stream's surface, nearly 18 meters (60 feet) down from the top of the ridge at the woodland site. The fossil layer ranges in thickness from 1.3 centimeters (0.5 inch) to approximately 15 centimeters (6 inches). By digging out the dirt, hauling it home in bags, and sifting every ounce, Denton and O'Neill have found teeth and bone fragments representing at least three Late Cretaceous dinosaurs, including *Dryptosaurus, Hadrosaurus,* and *Hypsibema.* Also present is fossil evidence of the giant crocodile *Deinosuchus.*

The Ellisdale site was recently incorporated into the Monmouth County, New Jersey, parks system, with management in perpetuity by the New Jersey State Museum. Denton and O'Neill are site managers and continue to visit after heavy rains.

Monmouth Brook Sites

The dinosaur fossil sites that are most accessible to amateur collectors in New Jersey are the Monmouth County brook localities. Several small streams near Marlboro, New Jersey, including Big Brook and Poricy Brook, expose Upper Cretaceous sediments of the Navesink and Mt. Laurel/Wenonah formations.

Although Parris, Gallagher, and other paleontologists have collected at the sites, amateurs have found most of the fossil treasures. Among the collectors have been high-school students from Sharon Hill, Pennsylvania. Led by Edward M. Lauginiger, a biology teacher at Academy Park High School, the students have collected nearly four thousand vertebrate fossils at Big Brook since 1980. The results of the collecting were published by Lauginiger in 1986.

The fossil inventory includes dinosaur remains, crocodile bones, turtle shell, shark teeth, and the remains of skates, rays, bony fishes, and mosasaurs. Ralph Johnson and others from the Monmouth Amateur Paleontologists Society have regularly scouted Big Brook, and he identifies at least three dinosaurs from the site, including *Dryptosaurus, Hadrosaurus,* and *Coelosaurus.* At Poricy Brook, bone fragments of nodosaurids—heavily armored dinosaurs—have reportedly been found.

New York

The Empire State yielded its first dinosaur fossils in 1972, when Paul Olsen and Robert F. Salvia discovered footprints at Nyack Beach State Park in Haverstraw, Rockland County, New York. The Late Triassic tracks (fig. 9.20) were found in the Newark Basin's Stockton Formation, which is also exposed in New Jersey and Pennsylvania. Rock slabs taken from the site and now at the New York State Museum reveal typical *Grallator* footprints. There are at least two crossing trackways, with individual prints ranging in size from 12 to 15 centimeters (5 to 6 inches). Another dinosaur footprint, possibly *Atreipus,* has been identified from these Stockton sediments.

No body fossils of dinosaurs have yet been found in New York, but skeletal evidence from Rockland County exists for a wide range of other reptiles. The fossils include phytosaur teeth, indeterminate reptile bones and scutes, and the archosaur footprints *Rhynchosauroides, Chirotherium, Brachychirotherium,* and *Apatopus.*

FIG. 9.20 *Grallator* footprints discovered in 1972 at Nyack Beach State Park in Rockland County, New York. (Published with permission of the Geological Survey/New York State Museum)

Delaware

Delaware has many of the same Upper Cretaceous formations that have produced an abundance of dinosaur fossils in New Jersey. Unfortunately, there are virtually no natural exposures of those sediments in the state. The only location in Delaware where dinosaur fossils have been found is along the Chesapeake and Delaware Canal, where eagle-eyed collectors have tri-

FIG. 9.21 Deep Cut site on the Chesapeake and Delaware Canal in Delaware, where dinosaur fossil–bearing sediments are now flooded.

umphed over conditions that have been compared to searching for a needle in a haystack.

Chesapeake and Delaware Canal

The 30 kilometer (19 mile) long Chesapeake and Delaware Canal connects the Chesapeake Bay with the Delaware River, angling across a belt of south-ward-trending Cretaceous formations. Originally built above sea level between 1824 and 1829 as a shortcut to the port of Baltimore, the canal was privately owned and operated with a system of water locks. In 1919 the federal government assumed ownership and began a series of dredging operations that deepened and widened the canal to make it a sea-level waterway for cargo ships. In recent times the U.S. Army Corps of Engineers, which operates the canal, has dredged periodically to maintain the shipping channel.

Cretaceous fossils were originally found in the 1820s in bluffs created by canal excavations. The site known as Deep Cut (fig. 9.21) exposed sediments of the Englishtown, Merchantville, and Marshalltown Formations and yielded abundant marine fossils, including ammonites, shark teeth, and fish bones. Deep Cut was flooded in 1982 by a canal stabilization project. Another historic locality that was lost to maintenance work was the Biggs Farm site, which produced more than 250 types of invertebrate fossils from the Mt. Laurel Formation.

174

FIG. 9.22 Dredge spoil pile at the Chesapeake and Delaware Canal. (Courtesy of Edward M. Lauginiger)

FIG. 9.23 Ed Lauginiger, a high-school biology teacher, explores a fossil site on the banks of the Chesapeake and Delaware Canal.

The only dinosaur fossils from Delaware have come to light in the "spoil" dredged from the canal and piled on the banks alongside (fig. 9.22). The main watchdog of dredge opportunities for fossil hunting is Ed Lauginiger. A resident of Wilmington, Delaware, Lauginiger (fig. 9.23) is an avid fossil hunter and tireless in advocating the joys of paleontology to his high-school biology students in Pennsylvania. His publications include "Delaware Fossils" (1981) with Eugene F. Hartstein, and *Cretaceous Fossils from the Chesapeake and Delaware Canal: A Guide for Students and Collectors.*

The inventory of canal dinosaur fossils is modest, all of it derived from Late Cretaceous Marshalltown and Merchantville sediments and none of it diagnostic enough for genus and species determination (fig. 9.24). At least two hadrosaurid teeth have been found at the canal, one by Lauginiger and another by Larry Decina, in spoil from a major Corps dredging project in the winter of 1984–1985. Other remains include several toe bones from ornithomimosaur dinosaurs and a partial hadrosaurid vertebra found in 1982 by Bill Gallagher of the New Jersey State Museum.

The spoil piles have also produced a rich assortment of marine fossils, including teeth of the seagoing reptiles *Mosasaurus, Globidens,* and *Tylosaurus;* a partial jaw and scutes of the giant crocodile *Deinosuchus;* a vertebra of a ple-

FIG. 9.24 Scouting the dredge spoil piles for fossils at the Chesapeake and Delaware Canal near the St. Georges Bridge. (Courtesy of Edward M. Lauginiger)

siosaur, remains of bony fishes, and shell fragments of the turtles *Trionyx, Tox-ochelys,* and several other forms. In 1981 Donald Baird and Peter M. Galton described remains of a pterosaur found in Merchantville Formation sediments dredged up in canal spoil piles.

Galton, professor of anatomy at the University of Bridgeport in Connecticut, has long been studying Late Triassic and Early Jurassic dinosaurs, particularly the prosauropods of the Connecticut Valley. The British-born paleontologist provided the most comprehensive research on North American prosauropods in 1976 and was the first to recognize the ornithischian nature of *Thecodontosaurus gibbidens* (chapter 5), which was renamed *Galtonia* in his honor by Hunt and Lucas in 1994.

Maryland

Maryland remains the only source of Early Cretaceous dinosaur fossils on the East Coast. The Arundel Clay sediments that yield these fossils are carefully watched by a handful of enthusiasts who monitor new exposures created by road cuts, building excavations, and quarrying operations in the heavily populated Baltimore-Washington corridor. Many local fossil hunters are affiliated with the Natural History Society of Maryland in Baltimore or the Maryland Geological Society in suburban Washington. Their diligence has paid off with important new additions to the Early Cretaceous dinosaur fossil record as well as rare finds from Upper Cretaceous marine sediments.

Muirkirk Arundel Clay

The area that produced the bulk of Maryland's dinosaur fossils in the last century is still the source of nearly all Early Cretaceous collecting in the state. Muirkirk—south of Laurel, in Prince Georges County—was home to the iron ore pits explored so fruitfully for fossils by John Bell Hatcher in 1887–1888 and Arthur Bibbins in 1894–1896 (chapter 4).

Today the best exposure of the Arundel is in a clay quarry at Beltsville operated for brick manufacturing (fig. 9.25) by Maryland Clay Products. Situated at the heart of the historic Muirkirk localities, the site has been commercially quarried for brick clay since 1939. Anecdotal evidence suggests that dinosaur teeth and bones have been found by quarry workers and amateur collectors over the years. Among the documented discoveries is a partial theropod limb bone found in 1974 by Peter Rath, plant manager at the clay quarry for more than forty years and now retired.

FIG. 9.25 Exposure of the Lower Cretaceous Arundel Clay at the Maryland Clay Products quarry near Beltsville in 1993.

FIG. 9.26 Peter Kranz, author of *Dinosaurs in Maryland,* examines a large iron ore concretion at the Maryland Clay Products site.

FIG. 9.27 Tom Lipka collecting Early Cretaceous plant material from an Arundel Clay site at Rossville, Maryland, in 1992. (Courtesy of Tom Lipka)

Systematic scientific collecting at the quarry was revived in the mid-1980s by Peter M. Kranz (fig. 9.26), who has a doctorate in geology from the University of Chicago. Kranz, a resident of Washington, D.C., is widely known in Maryland and the District of Columbia area as a popularizer of dinosaurs, and he conducts regular fossil-hunting field trips for the public. Kranz wrote the popular 1989 Maryland Geological Survey publication *Dinosaurs in Maryland.*

Since 1988 Kranz and his colleagues have added dozens of dinosaur fossils to the Maryland inventory, most of which is preserved at the Smithsonian Institution's National Museum of Natural History. His first confirmed dinosaur fossil from the quarry came in fall 1988, when he discovered a theropod toe bone. Other recent dinosaur finds at the site include teeth and limb bones from indeterminate theropods and from a sauropod dinosaur, probably *Astrodon.* Kranz also located a turtle skeleton in 1988 that was collected by Robert Weems of the U.S. Geological Survey. The skeleton represents the most complete Early Cretaceous vertebrate specimen ever brought to light in Maryland.

Since 1989 Thomas R. Lipka (fig. 9.27) has also regularly explored the Maryland Clay quarry exposures. Lipka, an electronics technician for the Baltimore City Department of Transit and Traffic, has collected abundant fossil material, largely teeth, representing Early Cretaceous theropods, sauropods, ankylosaurs, and crocodiles.

Three dinosaur finds in particular have raised hopes for further important discoveries at the Maryland quarry. In June 1990 Kranz unearthed a tooth that may represent a juvenile tenontosaur or possibly the only ceratopsian ever found in eastern North America (chapter 7).

The second major find occurred in May 1991 when Arnold Norden and his children, Heather and John, were prospecting the site. Norden, an aquatic ecologist with the Maryland Department of Natural Resources, happened upon the end of a huge *Astrodon* femur exposed by the clay quarrying operations. Carefully collected by the Smithsonian and put on display at the museum, the 1.2 meter (4 foot) fossil—part of an original femur estimated at 1.8 meters (6 feet) long—is the largest dinosaur bone yet discovered on the East Coast. Another partial *Astrodon* femur was collected by Robert Eberle in August 1989 from an exposure of the Arundel Clay near Arbutus in Baltimore County, Maryland.

The third significant discovery at the clay quarry was made by Kranz in May 1992, when he located the only associated dinosaur bones ever found

FIG. 9.28 Smithsonian specialists Dan Chaney (*left*) and Michael Brett-Surman collect dinosaur remains from the Arundel sediments at Maryland Clay Products in 1992.

in the Early Cretaceous of Maryland. Collected by specialists from the Smithsonian (fig. 9.28), the theropod material consists of the upper region of the tibia and an assortment of toe bones, ribs, and other skeletal remains (chapter 7).

The rich dinosaur heritage represented by the Arundel exposures in the Muirkirk area has prompted enthusiasts to seek creation of a state interpretive park on land adjacent to the quarry. Kranz has also persistently lobbied to have *Astrodon johnstoni* designated the official dinosaur of Maryland.

Severn Sites

The Upper Cretaceous Severn Formation is a narrow band of nearshore marine deposits that crops out on the eastern edge of Maryland's western shore. Collectors have discovered an abundance of teeth and bone fragments from marine reptiles such as mosasaurs, plesiosaurs, crocodiles, and turtles. A Kranz-led collecting foray in June 1989 to Severn exposures in Oxon Hill found the most complete mosasaur skeleton ever from Maryland.

Although dinosaur fossils from the Severn Formation are extremely rare, recent discoveries have been made in Prince Georges County east of Washington, D.C. Donald Baird has described the sites as "stream banks, road cuts, and excavations for construction" that are "small, scattered, and in some cases ephemeral."

Fragmentary remains of a hadrosaurid have been found in the Severn Formation, no great surprise since these dinosaurs were abundant in New Jersey and Delaware in the Late Cretaceous. Fragments of a hadrosaurid femur and small tibia were uncovered near Brightseat. In 1987 a hadrosaurid tooth was found in Bowie, Maryland. The other Severn dinosaur, *Coelosaurus,* has been tentatively identified from fossils collected from at least three sites in Prince Georges County: a partial left femur from the Bowie area, a small fragment of metatarsal from exposures near Largo, and a vertebra from Brightseat.

Pennsylvania

Dinosaur bones are extremely rare in Pennsylvania. The only known dinosaur skeletal remains are several teeth from Emigsville in York County, described by Edward Cope in 1878 as *Thecodontosaurus gibbidens* (now *Galtonia gibbidens*). However, many bones of other extinct vertebrates, such as phytosaurs and metoposaurs, have been found in the state.

The dinosaurs of Pennsylvania primarily left their mark in footprints, most discovered before the modern seekers began scouring the East Coast for new fossil evidence. But several new sites in southeastern Pennsylvania have yielded dinosaur tracks in recent years, augmenting the historical record.

Limerick

On November 9, 1981, construction was interrupted on the nuclear-powered Limerick Generating Station east of Pottstown, Pennsylvania, by the discovery of "horse prints" on a sandstone rock layer exposed during blasting for a spray pond. About 150 meters (500 feet) from the distinctive cooling towers at the $5 billion Philadelphia Electric Company project, workmen removed eighteen track-bearing slabs from approximately 7 meters (22 feet) below the ground surface (fig. 9.29).

Nearly sixty footprints from 4 to 15 centimeters (1.5 to 6 inches) long were imprinted on the sandstone slabs. They were identified in a December 1981

FIG. 9.29 Dinosaur footprint–bearing slabs uncovered during construction of Limerick Nuclear Generating Plant in 1981. (Philadelphia Electric Company)

report by one of the authors (DBW)—a University of Pennsylvania graduate student at the time of the discovery—as Late Triassic dinosaur tracks from the Lockatong Formation. The site is just west of Schwenksville in Montgomery County, where Earl Poole discovered reptile footprints, including dinosaur tracks, in 1939 (chapter 4).

At least two different reptiles are represented at Limerick by the 190 million year old prints. The most abundant are the three-toed theropod dinosaur prints known as *Grallator*. Nondinosaur tracks include the four-toed crocodile-like archosaur *Chirotherium*. The Philadelphia Electric Company, owner and operator of the power plant, donated most of the tracks to schools and museums. A single footprint-bearing slab is on display at the Limerick Generating Station visitors' center.

Coopersburg

In 1978 James A. Turner Jr. of Bethlehem, Pennsylvania, discovered fossil footprints in an outcrop of red shale near a parking lot southeast of Coopersburg in Lehigh County. Turner alerted the William Penn Museum (now the State Museum of Pennsylvania) in Harrisburg and the geology department at Lehigh University in Bethlehem. In the fall of 1978 Lehigh geology professor J. Donald Ryan and student volunteers excavated the Coopersburg site and uncovered a large number of additional reptile prints, ranging from 2 to 12 centimeters (0.8 to 5 inches) long. A number of trackways were also discovered.

Donald Baird later identified at least four types of reptile tracks from the nearly two-acre Upper Triassic outcrop at Coopersburg. They include the theropod track *Grallator;* the four-toed print of the phytosaur *Rutiodon;* and the primitive archosaurian tracks *Chirotherium* and *Rhynchosauroides brunswickii,* first described in 1947 by Ryan and Bradford Willard of Lehigh University from exposures at nearby Kintnersville.

Turner's plans to raise money for a museum at Coopersburg were unsuccessful, and the unprotected site was vandalized in June 1989. According to newspaper accounts, thieves used power saws to cut out 3 meters (10 feet) of slab bearing *Grallator* prints. The stolen tracks have never been recovered. In 1992 the site was reopened and additional *Grallator* tracks were collected by Peter Kranz and Tom Lipka (both from Maryland) and Robert Weems, a geologist with the U.S. Geological Survey in Virginia. Some of these Coopersburg prints are now part of the "Dinosaur Garden" display at the U.S. Botanic Garden in Washington, D.C.

Reading Area

In 1989 two fossil collectors from Pennside, a suburb of Reading in Berks County, made headlines in local newspapers with their discovery of reptile tracks in the Reading area. Michael J. Szajna and Brian W. Hartline (fig. 9.30) found their first print in Upper Triassic sediments at an excavation for a housing development in Exeter Township. Since then they have located tracks at dozens of small sites in Berks County and neighboring Montgomery County. Most of the localities have been temporary exposures created by construction work in Exeter Township, plus sites near the towns of Douglassville (east of Reading) and Sanatoga in Montgomery County.

In 1993 Szajna, president of the Berks Mineral Society, published the details of these track sites with Shaymaria M. Silvestri of the Department of Geological Sciences at Rutgers University. Among the Late Triassic dinosaur trackmakers they identified were two of the most common from that period on the East Coast, *Atreipus* and *Grallator*.

Most of the tracks consist of primitive archosaurian prints: *Apatopus, Batrachopus, Chirotherium, Gwyneddichnium,* and *Rhynchosauroides*. Szajna also reports the discovery of two new primitive crocodile-like tracks, as yet unnamed. Archosaur prints identified as *Brachychirotherium* are the youngest found so far in the Newark Basin, dating from just 2 million years before the Triassic-Jurassic boundary.

FIG. 9.30 Mike Szajna (*left*) and Brian Hartline display a *Grallator* footprint from the Reading, Pennsylvania, area. (Courtesy of Mildred Hartline)

FIG. 9.31 Dinosaur footprint horizon uncovered in 1993 at Graterford, Pennsylvania, on the grounds of the State Correctional Institution. (Courtesy of Michael Hendricks)

Graterford

In late 1993 a large dinosaur footprint site came to light on the grounds of the State Correctional Institution at Graterford in Montgomery County. Wayne Covington found the prints in an outcrop of the Upper Triassic Lockatong Formation exposed on the fenced-in prison property. More than one hundred three-toed footprints of various sizes are scattered on bedding planes in two ravines (fig. 9.31). Some are superimposed on fossilized mud cracks and ripple marks. Researchers from the State Museum of Pennsylvania (Robert M. Sullivan, Kesler Randall, and Michael Hendricks) and the Pennsylvania Geological Survey (William E. Kochanov) are now studying these tracks.

The researchers have identified at least three footprint species at Graterford. The three-toed dinosaur tracks range from 9–9.5 centimeters (3.5–3.7

inches) to 15 centimeters (6 inches) and represent *Grallator* and *Atreipus*. Smaller prints—five-toed and up to 2 centimeters (0.8 inch) long—are the archosaur tracks *Gwyneddichnium* or *Rhynchosauroides*.

Virginia

Unlike most other states on the East Coast, Virginia has yielded most of its dinosaur fossils in modern times. A growing inventory of footprints is testimony to the efforts of those who regularly scout the Upper Triassic and Lower Jurassic sediments in the state.

Culpeper

By far the most abundant dinosaur fossil discoveries in Virginia have been made in Stevensburg, east of Culpeper. The site is a stone and gravel quarry 300 feet deep (fig. 9.32) operated by Martin Marietta Aggregates on land owned by the Carpenter family. In the past twenty years, two separate horizons in the Balls Bluff Siltstone have been located and studied by researchers from the U.S. Geological Survey. The estimated 4,800 dinosaur and other reptile tracks at the quarry place it among the top sites worldwide with the largest number of reptile footprints.

Research on the first surface began in 1975. Robert E. Weems (fig. 9.33), then a doctoral student in geology at George Washington University, had been alerted to the presence of footprints by colleagues. Over the next two years, a one-acre track site was uncovered approximately 46 meters (150 feet) down from ground level and containing 830 reptile tracks in thirty-two trackways. The surface was covered with ripple marks and mud cracks that indicated the area was a mudflat along a lake margin approximately 215 Ma, during Late Triassic time. Weems—since 1978 a geologist with the U.S. Geological Survey in Reston, Virginia—summarized his Culpeper research in 1987. He identified a variety of dinosaur footprints, including *Grallator, Kayentapus, Gregaripus,* and *Agrestipus* (chapter 5). Among his many other researches into the ancient life and geology of Virginia, in 1980 Weems described an unusual archosaur found at a sewer plant construction site southeast of Doswell in 1974. He chose the name *Doswellia kaltenbachi* in honor of fellow graduate student James Kaltenbach, who made the discovery on a field trip with Weems and Nicholas Hotton III of the Smithsonian Institution.

Although the upper footprint surface at the Culpeper quarry represented a major discovery, there was more to come. In April 1989 quarry worker

FIG. 9.32 Stone and gravel quarry near Culpeper, Virginia, has yielded thousands of dinosaur and other reptile footprints.

FIG. 9.33 Rob Weems of the United States Geological Survey at the lower footprint site near Culpeper in 1992.

FIG. 9.34 Closeup of a dinosaur footprint from the lower level of the quarry near Culpeper.

Robert Clore spotted three-toed "bird tracks" during blasting some 76 meters (250 feet) below ground level. This lower surface—estimated to be 300,000 years older than the upper layer—is a 2 to 6 foot thick sheet of gray siltstone, tilted down six degrees to the west and exposed for approximately six acres.

Since 1989 Weems—assisted by U.S. Geological Survey colleagues Ronald J. Litwin and Nancy J. Durika—has found nearly 4,000 footprints on the lower bedding plane (fig. 9.34), including 2,300 *Kayentapus* tracks. Weems has also identified *Grallator* tracks and some made by aetosaurs. Trackways are abundant at the site (fig. 9.35), with the longest comprising 230 individual theropod footprints stretching 265 meters (865 feet). Litwin, who has a doctorate in geology from Pennsylvania State University, discovered that several of the *Kayentapus* trackways repeatedly cross those made by the aetosaurs, which means the latter may have been hunted by the dinosaurs.

Leaksville Junction

On the border of Virginia and North Carolina, what may be the oldest dinosaur footprints on the East Coast were uncovered in a mudstone quarry

FIG. 9.35 Dinosaur trackway at the quarry near Culpeper.

operated by the Solite Corporation. The quarry actually straddles the border, but its official address is Leaksville Junction, Virginia, near the town of Cascade. The tracks, discovered in the mid-1970s in sediments of the Upper Triassic Cow Branch Formation, are under study by Paul E. Olsen and Nicholas C. Fraser. A Ph.D. geologist from Aberdeen University in Scotland, Fraser is curator of vertebrate paleontology at the Virginia Natural History Museum in Martinsville. The quarry has yielded the dinosaur footprints *Atreipus* and *Grallator,* plus skeletal remains of the phytosaur *Rutiodon* and also *Tanytrachelos,* a small—20 to 40 centimeters (8 to 16 inches) long—lizardlike reptile.

Manassas Sites

In 1991 geologist Geoff Christie found three dozen dinosaur footprints at a construction site south of Manassas, Virginia, in the Upper Triassic Manassas Sandstone. The 220 million year old tracks, ranging up to 15 centimeters (6 inches) long, came from both theropods and ornithischians. These tracks currently are under study by the U.S. Geological Survey. In 1992 Weems found a single *Grallator* track in the Balls Bluff Siltstone during fieldwork along the banks of Bull Run Creek near Manassas National Battlefield. He has also recently identified a phytosaur femur from another Manassas area site, and local fossil hunter Mitch Berman found several well-preserved *Brachychirotherium* tracks in this formation near the same site.

Oak Hill

The dinosaur footprints and trackways at the Oak Hill estate in Aldie, Loudoun County (chapter 4), have attracted renewed attention from paleontologists reconstructing Virginia's dinosaur past. In 1985 Norma Kay Pannell, a master's degree student in geology at George Washington University, identified the theropod trackmakers *Grallator* and *Eubrontes* (chapter 6) at the site. A short trackway reveals a *Grallator*-like dinosaur, 53 centimeters (21 inches) high at the hip, trotting along at 1.8 meters a second (4 miles an hour) with a 100 centimeter (40 inch) stride. Dated at approximately 190 Ma, these Early Jurassic tracks are some 30 million years younger than the footprints in the quarry near Culpeper to the south.

North Carolina

Dinosaur fossil hunting in North Carolina has historically centered on Upper Cretaceous exposures of the Black Creek Formation along the Cape Fear

River in the southeastern part of the state. In addition, much older dinosaur remains from the Late Triassic have come from the Pekin Formation of the Deep River Basin in south-central North Carolina, near the Peedee River in Anson County. Two dinosaur teeth from the Pekin have recently been identified as belonging to the primitive ornithischian *Pekinosaurus olseni.*

Cape Fear River Sites

The most productive dinosaur fossil site in North Carolina history is at Phoebus Landing near Elizabethtown in Bladen County (fig. 9.36). The fossil-bearing exposure, reachable only by boat, is often submerged and has slumped considerably in recent years. Deposited in marine conditions some 75 Ma, Phoebus Landing contains a diverse mix of scrappy marine and terrestrial vertebrate fossils from the Late Cretaceous.

Although Cretaceous vertebrate fossils were described from the Cape Fear River Basin as early as 1858 by Ebenezer Emmons, it was in 1905 that collecting of dinosaur and other reptile bones was initiated at Phoebus Landing (fig. 9.37) by Lloyd W. Stephenson and Edward W. Berry. Beginning in 1963, Halsey W. Miller Jr.—then on the faculty at nearby High Point College—

FIG. 9.36 Phoebus Landing dinosaur fossil site on the Cape Fear River near Elizabethtown, North Carolina. (New Jersey State Museum)

FIG. 9.37 Collecting at the Upper Cretaceous Phoebus Landing site in North Carolina. (New Jersey State Museum)

greatly expanded the dinosaur inventory at Phoebus Landing. Miller published revised faunal lists in 1967 and 1968, which were updated in 1979 by Donald Baird and Jack Horner.

Among the discoveries at Phoebus Landing are remains of the theropods *Dryptosaurus* and *Coelosaurus,* the hadrosaurid *Hypsibema,* indeterminate hadrosaurids, the giant crocodile *Deinosuchus,* and the mosasaur *Tylosaurus.* A diverse fish fauna is also present. Phoebus Landing is similar in depositional environment to the Ellisdale site in New Jersey, and the New Jersey State Museum conducted scientific expeditions at the North Carolina location in 1986 and 1989. The site similarities were documented in a 1986 publication by Bill Gallagher and his museum colleagues.

Although the vast majority of fossil collecting in North Carolina involves Eocene through Pleistocene exposures, a handful of dedicated collectors from fossil clubs in the state and nearby South Carolina monitor the Upper Cretaceous sites. The Durham-based North Carolina Fossil Club fields a number of Cretaceous dinosaur hunters, including Tom Burns of Sanford. Burns reports productive sites throughout the Cape Fear River valley, from Tarheel south into Brunswick County and along many of the river's tributaries. He has found numerous hadrosaurid teeth, vertebrae, and limb and toe bones, plus rare theropod teeth. In 1985 Burns located fossil evidence at Phoebus Landing of an ornithomimosaur. Other active collectors in the North Carolina Cretaceous include Dwayne Varnum and Frank and Becky Hyne.

Several collectors from across the border in South Carolina have also hit paydirt at Phoebus Landing. Aura Baker of the Myrtle Beach Fossil Club and Rita McDaniels of the Grand Strand Fossil Club in Myrtle Beach have reportedly found hadrosaurid and theropod material at the North Carolina site.

Some of the amateur collectors work in collaboration with the North Carolina State Museum of Natural Sciences in Raleigh. Vince Schneider, a paleontology research associate at the museum, reports the discovery of a partial hadrosaurid femur from Goldsboro in Wayne County about 1980, plus a small theropod tooth from a river site near Phoebus Landing in 1990. Additional undescribed dinosaur remains from North Carolina are in amateur collections.

South Carolina

Until the mid-1980s, no dinosaur fossils had been found in South Carolina. The same Upper Cretaceous sediments that yielded dinosaur material in North Carolina were known to run through the state, but collectors had come up empty-handed. That began to change in 1986 when Aura Baker, president of the Myrtle Beach Fossil Club, found two hadrosaurid teeth from a site near Kingstree in Williamsburg County. At about the same time, a high-school student found a theropod tooth in a sand pit near Quinby in Florence County.

These Late Cretaceous teeth constituted the entire South Carolina dinosaur fossil record until the early 1990s, when several new sites produced a small but tantalizing inventory of indeterminate dinosaur material. The discoveries have been studied and cataloged by James L. Knight, curator of natural history at the South Carolina State Museum in Columbia, and most are now on display at the museum. The new sites are all in the Upper Cretaceous Donoho Creek Formation (formerly Black Creek Formation) and have been collected by a Quinby-based fossil club, the Midstate Geological Research Team. Leaders of the group include Derwin Hudson, Lee Hudson, and Ray Ogilvie.

FIG. 9.38 Burches Ferry fossil site in South Carolina. (James L. Knight, South Carolina State Museum)

Burches Ferry Area

This site in Florence County is seldom exposed because of a strictly controlled flow schedule from dams along the Peedee River. The first dinosaur fossil discoveries at the site (fig. 9.38) were two fragments of the same theropod tooth, found several weeks apart by different collectors. Diligent searching by Lee Hudson has since produced a theropod claw.

Lynchburg

A locality at Lynchburg on the Lynches River southwest of Florence yielded a whole theropod tooth in 1991, in spoil from a paving company pit.

Quinby

At least three sites near Quinby have yielded single theropod or hadrosaurid teeth, and part of a theropod metatarsal, or foot bone, was found by

Derwin Hudson in 1992. Work at Quinby and elsewhere in the state's Upper Cretaceous sediments continues with the hope of finding more evidence of these extremely elusive South Carolina dinosaurs.

Nova Scotia

The system of Newark Supergroup basins extends all the way north into the Canadian maritime province of Nova Scotia, where 15 meter (50 foot) tides in the Bay of Fundy have exposed Triassic-Jurassic sediments of the Fundy Group along the shores of the Minas Basin. The 240 km (150 mile) Bay of Fundy divides Nova Scotia from New Brunswick.

Since the mid-1980s the Fundy area has yielded a bonanza of dinosaur and other reptilian fossils, as well as the near relatives of mammals. This recent surge of discovery is the second "bone rush" in Nova Scotia. The first occurred in 1851 at Joggins, on the south shore of the Cumberland Basin. Sir William Dawson, a native of Nova Scotia, and the Scottish geologist Sir Charles Lyell discovered the remains of the world's first terrestrial reptile among the 300 million year old fossil tree trunks at Joggins.

Upper Triassic Sites

Modern dinosaur hunting in Nova Scotia was pioneered by Donald Baird, who began prospecting the northern sites in the mid-1960s. Later he was assisted by Jack Horner, a preparator at the Princeton museum in the late 1970s, and Paul Olsen, who began traveling north with Baird during summer breaks while he was still in high school in New Jersey.

Among the discoveries are dinosaur fossils from the Upper Triassic Wolfville Formation in Kings and Hants counties. In addition to Baird, Horner, and Olsen, researchers in the Wolfville sediments include Hans-Dieter Sues, associate curator of vertebrate paleontology at the Royal Ontario Museum in Toronto. Sues (fig. 9.39), born in Rheydt, a small town near Dusseldorf, Germany, earned his Ph.D. in biology from Harvard University. His other major East Coast discoveries (with Olsen) include a rare collection of 225 million year old remains of cynodonts—mammal-like reptiles—found at a site near Richmond, Virginia, in 1989.

From the cliffs and beach southeast of Paddy's Island in Kings County, the Wolfville Formation has produced *Atreipus* footprints and trackways plus the archosaurian tracks *Rhynchosauroides* and *Brachychirotherium*. Dinosaur skeletal remains from the Wolfville include an ornithischian tooth and some pro-

FIG. 9.39 Hans-Dieter Sues of the Royal Ontario Museum in the field in 1985 near Parrsboro, Nova Scotia. (Courtesy of Hans-Dieter Sues, Royal Ontario Museum)

sauropod bones. Nondinosaur remains represent a small plant-eating reptile (possibly *Hypsognathus*) and an aetosaur. A partial metoposaur skull was found in 1974 at Noel Head in Hants County.

The Upper Triassic unit overlying the Wolfville, known as the Blomidon Formation, has been collected by Olsen and others at sites that include the north shore of St. Mary's Bay near Rossway in Digby County. This formation has produced *Grallator* footprints and an assortment of nondinosaur fossils such as *Rhynchosauroides* tracks and a snout fragment of the phytosaur *Rutiodon*.

Lower Jurassic Sites

The most famous and productive dinosaur fossil sites in Nova Scotia are on the north shore of the Minas Basin, near the town of Parrsboro. A tantalizing early discovery came in April 1984, when local fossil collector Eldon George found a 35 by 40 centimeter (14 by 16 inch) sandstone slab crisscrossed by five tiny trackways at Wasson Bluff (fig. 9.40) in the Lower Juras-

FIG. 9.40 Wasson Bluff, near Parrsboro, Nova Scotia, a major fossil site in the Fundy Basin. (Courtesy of Hans-Dieter Sues, Royal Ontario Museum)

FIG. 9.41 Neil Shubin (*foreground*) and Paul Olsen at Wasson Bluff in Nova Scotia. (Courtesy of Paula Chandoha)

sic McCoy Brook Formation. The footprints, no larger than a penny, were identified by Olsen as *Grallator*-like, and they have since been attributed to the world's smallest dinosaur, perhaps the size of a baby sparrow.

Wasson Bluff is the westernmost of a half-dozen neighboring fossil sites explored during the past twenty years by researchers including Olsen, Sues, Baird, Horner, Robert Grantham of the Nova Scotia Museum, and Neil H. Shubin (fig. 9.41), an assistant professor of biology at the University of Pennsylvania. Shubin has a Ph.D. in developmental biology from Harvard University. His specialty is the development and phylogeny of many groups of animals, including dinosaurs.

Hundred-foot-high cliffs stretch for miles along the Minas Basin shorefront: sandstone and basalt exposures that have yielded a rich assemblage of Early Jurassic dinosaur fossils. The sites include Blue Sac and McKay Head (Cumberland County) and Five Islands Provincial Park (Colchester County). The Scots Bay Formation, a lateral equivalent to the McCoy Brook Formation, has also produced dinosaur footprints.

In 1983 Shubin found fossil evidence at Wasson Bluff for tritheledonts, tiny animals thought to be the closest extinct relatives of mammals. The true bone rush at the site started in the summer of 1984 when Shubin, Olsen, and Sues began accumulating thousands of fragmented fossils from the beach and cliff faces, dating from near the Triassic-Jurassic boundary. The dinosaur material includes the footprints *Grallator, Eubrontes,* and *Anomoepus,* plus skeletal remains of the prosauropods *Ammosaurus* and *Anchisaurus* and yet to be described primitive ornithischians documented by isolated teeth and bones.

10

DINOSAUR MYSTERIES

Since the late 1960s, our view of dinosaurs has changed drastically. No more do we think of gigantic sauropods as stuck in the swamp, large theropods as inept carnivores, and the rest of the dinosaurian menagerie as pea-brained social misfits given to a lethargic life of eating and sleeping.

Most dinosaurs had sophisticated and complex social behavior, ranging from territorial and mating displays to organized herding. It's clear that some dinosaur parents cared for their young. And we now know that not all of them died at the sudden end of the Mesozoic: dinosaurs live on as birds.

How did this changing view of dinosaur biology come about? Have these new ideas put old controversies to rest? What new controversies have emerged? In answer to these questions, we will look at three "hot" issues that helped forge our new picture of dinosaurs and that still stand as major research efforts: birds as dinosaurs, dinosaur thermoregulation, and dinosaur extinction.

But first let's trace the roots of the dinosaur family tree (chapter 1) back to the ancestors of these great Mesozoic beasts.

Dinosaurs as Archosaurs

To identify the group that gave rise to dinosaurs, scientists compare the anatomy of animals like *Deinonychus, Tyrannosaurus, Brachiosaurus, Protocer-*

FIG. 10.1 *Lagosuchus,* a small meat eater from the Middle Triassic of Argentina, bears the closest relation to the dinosaurs of any archosaurs.

atops, and *Parasaurolophus* with as many other kinds of animals, living and extinct, as possible. The goal is to identify special similarities that reveal closeness of evolutionary relationships.

Crocodilians and birds share with dinosaurs a prominent opening in the side of the face (the antorbital opening) and similar braincase anatomy. In fact all three—dinosaurs, birds, and crocodilians—share numerous features as members of the group known as Archosauria ("ruling reptiles"). As archosaurs, crocodilians and birds are the closest modern relatives of dinosaurs. Crocodiles, both living and extinct, can be thought of as dinosaurs' second cousins. Yet these aquatic, sprawling, four-legged predators are highly modified from their common ancestor with dinosaurs. In fact some scientists think the earliest crocodiles were sleek and fast enough to gallop after their prey.

As recently as the mid-1980s, a group of archosaurs called thecodonts ("socket teeth") was thought to have given rise to dinosaurs as well as to crocodilians, birds, and pterosaurs. This is no longer considered true. In fact, Thecodontia as a group no longer exists, because some of these animals are more closely related to crocodilians, some to dinosaurs, and some to the entire archosaurian group.

If there is no evolutionary or genealogical "glue" holding a group together, then we must take a closer look at relationships between individual members. It turns out that pterosaurs, the flying reptiles of the Mesozoic, have a greater affinity with dinosaurs than do crocodilians. Pterosaurs are not dinosaurs, but they were dinosaurs' first cousins. From pterosaurs, it is a short jump by way of animals like the Middle Jurassic *Lagosuchus* ("rabbit crocodile," from Argentina) to the common ancestor of all dinosaurs. *Lagosuchus* (fig. 10.1) provides a glimpse of what this common ancestor may have been like: 1 meter (3 feet) or so long, walking and running on its hind legs, and agile and predatory in its eating habits.

From this diminutive ancestor came the great evolutionary radiation of dinosaurs, producing the small killer *Coelophysis* as well as the horrible and huge *Tyrannosaurus,* the super gigantic *Seismosaurus,* the ornithischians *Iguanodon, Dacentrurus, Pinacosaurus, Pachycephalosaurus,* and *Chasmosaurus*—and more!

Birds as Dinosaurs?

In 1861 the scientific community was rocked by the discovery of a spectacular fossil from the Upper Jurassic limestones of Bavaria. The skeletal remains were exquisitely preserved down to the impressions of downy feathers on its body and flight feathers on its arms and tail. This feathered animal was called *Archaeopteryx* ("ancient wing"), and it was clearly the first known bird (fig. 10.2). Since then, six additional *Archaeopteryx* skeletons have been uncovered.

FIG. 10.2 *Archaeopteryx,* the most primitive known bird, from the Late Jurassic of Germany.

From the very beginning researchers admired the anatomical detail of *Ar-chaeopteryx* and puzzled over its affinities. Two years after the initial discovery, British anatomist William K. Parker eloquently framed the significance of this new fossil:

> There is a curious blending of the characters of the various reptilian groups in the Birds. There has been no exclusive adoption of the mode of structure of any one scaly type by these feathered vertebrates. Those reptilian qualities and excellencies which are best and highest have become theirs. But how much more! This exaltation of the "Sauropsidan" or oviparous type by the substitution of feathers for scales, wings for paws, warm blood for cold, intelligence for stupidity, and what is lovely instead of loathsomeness—this detailed sudden glorification of the vertebrate form is one of the great wonders of Nature.

T. H. Huxley, "Darwin's bulldog," was the first to attempt to classify *Archaeopteryx* and identify its direct relatives. Huxley favored a dinosaurian ancestry for this first feathered reptile. For his hypothesis, he drew extensively on Darwin's theory of evolution—particularly the theory of common descent—to connect *Archaeopteryx* and its contemporary, the small theropod dinosaur called *Compsognathus* ("pretty jaw"). These two fitted beautifully as missing links between larger dinosaurs and modern birds, in large part because of their birdlike pelvis and hind limbs.

Huxley's ideas about birds and dinosaurs held sway until the 1920s, when the Danish anatomist Gerhard Heilmann wrote *The Origin of Birds* (1927). Many regard this study as the most comprehensive study of early avian evolution. Heilmann's aim was to document "with absolute certainty . . . that the birds have descended from the reptiles." He used his extensive knowledge of anatomy, embryology, and paleontology to speculate on which reptile group gave rise to birds.

Like Huxley, Heilmann at first considered small theropod dinosaurs like *Compsognathus, Ornitholestes,* and *Ornithomimus* to be close relatives of birds based on the large number of detailed similarities in their skeletons. "It would seem a rather obvious conclusion that it is amongst the Coelurosaurs that we are to look for the bird-ancestor," Heilmann wrote.

But he noted that the clavicles, or collarbones, of dinosaurs are absent in coelurosaurs. Clavicles are skeletal features also present in humans, separating the shoulder blades from the breastbone. Since Heilmann regarded evolution as strictly irreversible—what is lost cannot be regained—he concluded that "these saurians [dinosaurs] could not possibly be the ancestors of the birds."

Instead Heilmann came to focus on a newly discovered Triassic archosaur called *Euparkeria,* then classified as a pseudosuchian thecodont. What impressed him most was that these pseudosuchians had not progressed very far toward dinosaurs and therefore could have given rise to both birds and dinosaurs. What's more, they possessed clavicles, which are also found fused in birds to form the wishbone, the prize of the Thanksgiving turkey dinner.

But dinosaurs like *Compsognathus, Allosaurus,* and *Tyrannosaurus* don't have these special bones. In this way Heilmann regarded pseudosuchian thecodonts as the group having very special evolutionary relationships to birds. "Nothing in their structure militates against the view that one of them might have been the ancestor of the birds," he wrote.

It took nearly fifty years for Heilmann's ideas on avian origins to be seriously challenged, so strong was his argument on the pseudosuchian ancestry of birds. The new view, closer to Huxley's original ideas, came from the pioneering work of John H. Ostrom of Yale University (chapter 9) in the late 1960s. It was Ostrom who discovered *Deinonychus,* a 2 meter (7 foot) long theropod dinosaur from the Early Cretaceous of Montana and Wyoming. This small, fleet carnivore—an intelligent and adept hunter with a scimitar claw on each foot, possibly warm-blooded, remarkably birdlike—prompted Ostrom to reevaluate the evolutionary connection between *Archaeopteryx* and theropods. His research invigorated and inspired a great resurgence of interest in the origin of birds, so that today they are one of the most analyzed of the major vertebrate groups.

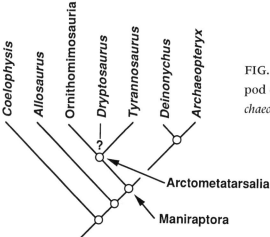

FIG. 10.3 Evolutionary tree of theropod dinosaurs, from *Coelophysis* to *Archaeopterx.*

With the discovery of *Deinonychus* and restudies of *Archaeopteryx,* it became clear that these two animals shared many special features with other theropod dinosaurs but not with any other group of archosaurs. For example, except for the flight plumage, the skeleton of *Archaeopteryx* is in many ways very similar to a number of small theropods like *Deinonychus, Coelophysis,* or *Compsognathus.*

Through these new studies, it's much easier to look at the "family" tree of early birds and their evolutionary connections with particular theropod dinosaurs. Figure 10.3 shows this evolutionary tree from *Coelophysis*—which itself may be closely related to our East Coast *Podokesaurus* and the makers of *Grallator* and *Eubrontes* tracks—through *Allosaurus* and several other more advanced theropods to *Archaeopteryx.* Note that some of these are given special names, such as Maniraptora: theropods with a highly flexed neck, long forelimbs, and an elongate middle finger.

Do birds have close relationships with some of these maniraptorans by way of *Archaeopteryx?* Yes. As the diagram shows, *Archaeopteryx* is most closely related to *Deinonychus.* All birds (including *Archaeopteryx*) and *Deinonychus* together are closely related to other theropods. For example, they are like cousins to Arctometatarsalia, a group of theropods introduced in chapter 8. A few of these arctometatarsalians are found along the East Coast: *Dryptosaurus aquilunguis* and ornithomimosaurs like *"Ornithomimus" affinis* and *Coelosaurus antiquus.*

TABLE 10.1   Late Cretaceous Avian Dinosaurs of the East Coast

Dinosaur	Locality	Geologic Formation	Description[a]
Saurischia			
Telmatornis affinis	New Jersey	Navesink and Hornerstown formations	S, small shorebird
Telmatornis priscus	New Jersey	Navesink Formation	S, small shorebird
Anatalavis rex	New Jersey	Hornerstown Formation	S, small shorebird
Graculavus velox	New Jersey	Hornerstown Formation	S, small shorebird
Laornis edvardsianus	New Jersey	Hornerstown Formation	S, small shorebird
Palaeotringa littoralis	New Jersey	Hornerstown Formation	S, small shorebird
Titthostonyx glauconiticus	New Jersey	Hornerstown Formation	S, small shorebird

[a]S = skeletal remains; F = footprints.

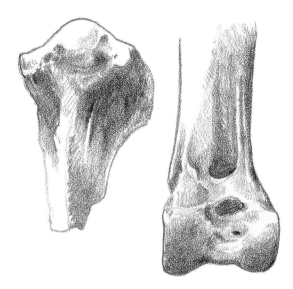

FIG. 10.4 Late Cretaceous birds from New Jersey: (*left*) top of the humerus (upper arm bone) of *Graculavus*; (*right*) lower end of the tibiotarsus (part of the lower hind leg) of *Laornis* (after Olson and Parris 1987).

So never mind Heilmann's thecodonts, a group that has no evolutionary glue holding it together. And never mind clavicles: there are so many similarities between the likes of *Deinonychus* and other maniraptorans that the comings and goings of clavicles really become a very small issue.

The recognition of birds as dinosaurs has an immediate impact on our East Coast bestiary, for there is a modest record of birds from New Jersey during the Late Cretaceous (table 10.1). The first of this record—from the Navesink and basal Hornerstown Formations (chapter 8)—includes a number of small shorebirds, originally recognized by Othniel C. Marsh in the late 1800s and most recently studied by Storrs L. Olson (Smithsonian Institution) and David C. Parris (New Jersey State Museum) in 1987. Known only from fragmentary wing and leg bones, these birds include *Graculavus* ("grackle wing"), *Telmatornis* ("marsh bird"), *Anatalavis* ("duck-winged bird"), *Laornis* ("stone bird"), and *Titthostonyx* ("little sharp point"), as well as others (fig. 10.4).

It's clear from these and other Mesozoic forms that birds *are* theropod dinosaurs, not just the closest relatives of theropod dinosaurs. Pigeons in city parks, ostriches on the African plain, migratory ducks and geese—all are living descendants of these extinct reptiles.

Origin of Flight

The advent of avian flight is as fascinating as the origin of birds itself. At present there are two opposing views. The "arboreal" theory suggests that bird flight developed from an initial stage of gliding down from trees. The ancestor of *Archaeopteryx* and all other birds was thought to have been quadrupedal, able to clamber up trees and leap and glide from branch to branch. *Archaeopteryx* was the second stage in the emergence of full avian flight. With its perching foot and fingers that remain free of each other, it certainly could have grasped limbs and perhaps glided through the treetops, flapping its flight-feathered wings only when necessary.

The other theory of flight origins—known as the "cursorial" theory—has our bird ancestor on the ground. Beginning with small, fast-running theropods like *Deinonychus,* flight supposedly evolved from successively higher and longer jumps into the air. In this way flapping flight took over from leaping as the animal attempted to overcome gravity for longer and longer periods. *Archaeopteryx* may thus represent the transition from a fast-running, ground-dwelling theropod to birds with fully powered flight.

There are pluses and minuses to both theories. The cursorial theory best embodies the now proven view that birds evolved from small, fast-running theropods. Yet many scientists have trouble imagining this scenario leading to successful flight. They argue that using gravity from the trees down would be much easier than fighting it from the ground up. However, no one has yet found fossil evidence to support the tree-dwelling theory for the predecessors of *Archaeopteryx.*

So we are stuck between function (what seems most likely when we consider the limits of gravity) and phylogeny (the evolutionary relationships for which we have direct evidence in the form of fossils). Given the intense interest in the origin of birds and their flight, we will undoubtedly be hearing about this debate for years to come.

Dinosaurs: Warm-Blooded or Cold-Blooded?

This question has fascinated the public and vexed many scientists for the past twenty-five years. Like the origin of bird flight, dinosaur thermoregulation is a complicated and often frustrating matter. James O. Farlow of Indiana University–Purdue University at Fort Wayne, one of the most active researchers in this field, has suggested that we may never know for sure whether dinosaurs were warm-blooded or cold-blooded.

Even so, there is some agreement among researchers that the dinosaurs most likely to have been warm-blooded (endothermic) are the small, agile, and deadly theropods. Those thought to have been cold-blooded (ectothermic) are the immense sauropods.

Early workers had no doubt that all dinosaurs were cold-blooded. After all, they were reptiles, weren't they? We should expect a physiology similar to that of snakes, lizards, and crocodiles. This means a generally slow pace of life and fairly simple behavior and interactions with other animals. Cold-blooded organisms take advantage of environmental heat sources like the sun to stay warm, a "cheap" way of thermoregulating because it requires very little expenditure of energy.

This view of dinosaurs began changing in the late 1960s, particularly with the work of Robert T. Bakker, first during his undergraduate years at Yale University and later at Harvard University (where he received his Ph.D.) in the 1970s, at the Johns Hopkins University in the early 1980s, and now in Colorado.

Nearly single-handedly, Bakker transformed dinosaurs from slow and sluggish swamp dwellers to lyrically leaping landlubbers. The key to this startling new view, Bakker concluded, was that dinosaurs were warm-blooded. Endothermy—an internal and costly way to produce heat—is also found in modern mammals and birds.

Bakker's perspective on dinosaur physiology is rooted in the origins of mammals and dinosaurs and their later history in the Mesozoic. Both groups evolved at the same time and more or less in the same neighborhood. Yet it was the dinosaurs that held sway for 165 million years, banishing mammals to small size and existence on the fringes of terrestrial habitats. This competitive superiority ceased with the great extinction 65 Ma, when the mammals finally got their chance to radiate into the many large and diverse forms we see throughout the Cenozoic Era, the "age of mammals" (chapter 2).

It's hard to imagine how the older view of dinosaurs as sluggish and stupid ever explained the dominance of dinosaurs over mammals during the Mesozoic. Nevertheless, we need evidence to try to establish warm-bloodedness in dinosaurs. Some of it comes from the fossils themselves, particularly in terms of anatomy. Other evidence is derived from the time and place where the fossils were found.

Walking, Breathing, and Warm-Bloodedness

In earlier chapters we talked a great deal about dinosaurs' locomotion and how their limbs were held fully erect under the body. This arrangement makes sense for large, land-living animals because it offers stronger and more stable support. All dinosaurs had this limb posture, as revealed by their bones and trackways. So do modern mammals and birds. In contrast, all other mod-

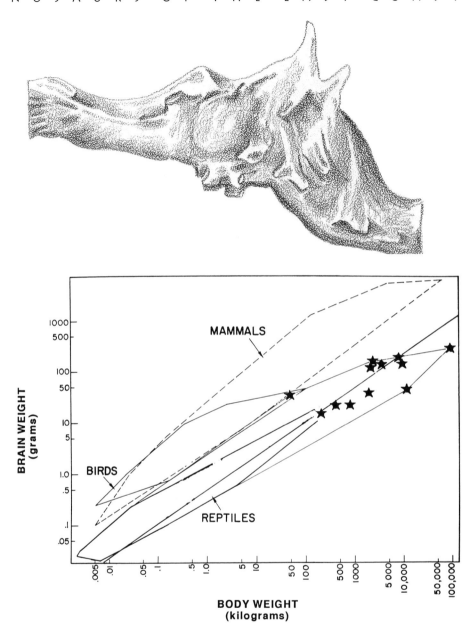

FIG. 10.5 Dinosaur brains: (*top*) the cast of the brain cavity of *Tyrannosaurus;* (*bottom*) relation between body size and brain size in dinosaurs compared with living birds, mammals, and reptiles (after Hopson 1977). Note that most dinosaurs fall within the expected values of reptiles scaled up to dinosaur size.

ern terrestrial vertebrates hold their legs out somewhat from the sides of the body: crocodilians semierect, and lizards, turtles, and amphibians in a sprawling posture.

Looking more closely, we see it's the warm-blooded groups that have the fully erect limb posture, while the cold-blooded bunch doesn't. Does that prove dinosaurs were endothermic? Not necessarily. In recent years David L. Carrier of Brown University has theorized that fully erect posture evolved to provide a better means of breathing. In lizards, crocodiles, and other forms with sprawling or semierect posture, the body becomes squished from side to side and the lungs can't expand and contract as efficiently as in animals like birds and mammals. Therefore fully erect limbs were a prerequisite for warm-blooded animals, which needed efficient breathing to support their greatly increased metabolic demands.

In keeping with the theme of endothermy's cost, it also makes sense that many dinosaurs were sophisticated chewers. Just look at the complexity of the jaws and teeth of ornithopods like *Hadrosaurus foulkii* (chapter 8) and other hadrosaurids. These dinosaurs even had a secondary palate so they could eat and breathe at the same time. Or look at the slicing-and-dicing equipment of ceratopsians like *Triceratops*. And the great predators of the time—*Tyrannosaurus, Deinonychus, Troodon*—had jaws well designed for tearing flesh from victims. All these complex jaws and teeth tell us their owners ate lots of food, perhaps to stoke the costly internal fires of endothermy.

What do dinosaur brains tell us about this debate on warm-bloodedness versus cold-bloodedness? James A. Hopson of the University of Chicago has argued that brain size is a function of the sum of all an animal's activities (fig. 10.5). If this is true, then sauropods, ankylosaurs, and stegosaurs—dinosaurs with quite small brains—appear to have had behavior levels and metabolic rates comparable to those of modern lizards and crocodilians. Ornithopods and large theropods (dinosaurs of larger brain size) were more sophisticated in behavior and probably had higher metabolic rates. Finally, the revved-up small theropods like *Deinonychus* and the ornithomimosaurs may have been as big brained as many living birds and mammals. Hopson regarded these latter dinosaurs as likely endotherms.

Bone Structure, Growth Rates, and Warm-Bloodedness

One of the most promising areas of research on dinosaur physiology focuses on the internal structure of fossil bone. Paleontologists have long known that fossil bone, when cut into thin sections and examined under a microscope, often resembles modern bone. There are channels for blood vessels, small holes for cells, rings that indicate growth rates (just like tree rings), and other features that can be seen in bone from animals living today.

All these features are important in understanding how bone grows. Early in bone development, special cells build what is called primary bone, of which there are two kinds. In "lamellar-zonal" primary bone, the bone-forming cells take their time, laying down solid sections of bone at such a slow rate that changing seasons can alter the pattern of deposition and produce growth rings. In contrast, "fibrolamellar" primary bone is made in a hurry by bone-forming cells (fig. 10.6). The basic internal form of this bone looks something like a scaffold, telling us that it was rapidly laid down and later filled in with secondary bone, often in the form of "haversian" bone.

It is the presence of fibrolamellar bone—which indicates rapid growth—that paleontologists are now targeting in their studies of the relationship between bone architecture and dinosaurian physiology. Fibrolamellar bone in dinosaurs seems to indicate that these animals had high growth rates, as high

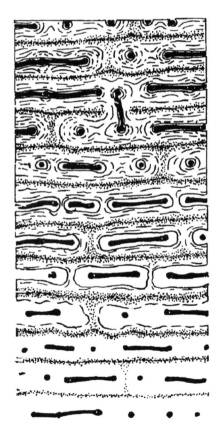

FIG. 10.6 Microscopic organization of fibrolamellar bone, the kind thought to characterize most if not all dinosaurs and also thought to indicate rapid growth rates (after Ricqlès 1980).

as or higher than those of modern birds and mammals. If so, then it may be that endothermy is necessary to sustain such rapid growth rates.

Any close physiological similarity between modern birds and the likes of *Tyrannosaurus, Brachiosaurus,* or *Triceratops* is due to their common ancestry, so these present-day endotherms might provide clues about the evolution of avian warm-bloodedness. We think all extinct birds were endothermic, especially since the most primitive form, *Archaeopteryx,* was covered with insulating feathers, obviously important in retaining heat if you're an endotherm but a nuisance if you're an ectotherm.

Can avian endothermy be pushed back down the theropod evolutionary tree? So far not yet, principally because of shortcomings in the fossil record. But if we were to find a specimen of *Deinonychus* or an ornithomimosaur that gave us just the right kind of detail—as the presence of feathers in *Archaeopteryx* has done for bird prehistory—then things could quickly change.

How Many and Where Found

There are other clues to dinosaur warm-bloodedness or cold-bloodedness. Some are provided through paleoecology—the study of ancient ecology, such as who were predators and who were prey—and some by paleobiogeography, the study of where dinosaurs and other extinct organisms have been found.

Bakker proposed what he called predator/prey biomass ratios (fig. 10.7). Simply put, endothermic animals have to eat far more food than ectothermic animals, so there should be far fewer warm-blooded than cold-blooded predators for a given amount of prey. In his calculations, Bakker factored in abundance of individuals, the total of their body masses, and even the sum of energy each group might have generated. He found that endothermic predators require an order of magnitude (that is, ten times) more "meat on the hoof" than ectothermic predators just to stoke their internal engines. In contrast, ectothermic predators don't need as much food because they rely on solar energy to keep warm.

Ingenious though such an approach is—and it gets to the heart of the cost of endothermy—there have been few advocates of Bakker's predator/prey biomass ratios. In the end there appear to be too many loose ends to provide great confidence in the method, and it has fallen by the wayside in studies of dinosaurian thermal physiology.

Finally, the geographic location of dinosaur finds may suggest how well these animals tolerated extremes in temperature, particularly high-latitude

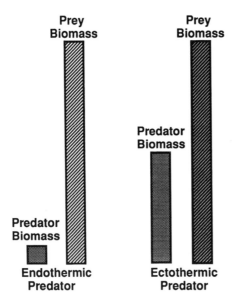

FIG. 10.7 Proportions of the mass of all predators to the mass of prey needed to feed them for both endothermic and ectothermic predators. This is what is known as predator/prey biomass ratios.

cold. The high latitudes during the Mesozoic were not as severe as they are today, with no polar ice caps from the Triassic through the Cretaceous and global climates much more equable than now. Still, there is evidence of freezing temperatures during the Cretaceous within the Antarctic Circle (Victoria, Australia) and the Arctic Circle (on the North Slope of Alaska). And each of those locations has yielded a good dinosaur fossil record.

Did these dinosaurs winter over in Victoria and Alaska, relying on their internal furnaces to keep them warm? How did they find enough to eat, since high latitudes suffer from six months of darkness? Did they migrate in and out of these polar regions with the growing seasons? These questions are under study by many scientists, among them dinosaur paleontologists, paleobotanists, geologists, and paleoclimatologists.

The mystery of dinosaur thermal biology so far remains unsolved. But with all the fascinating research going on in the field and laboratory, the answer may be just around the corner.

Dinosaur Extinction

Something drastic happened about 65 Ma, and the likes of *Tyrannosaurus, Velociraptor*, duckbills galore, and *Triceratops* all died out, whereas birds survived. Many other groups also perished, including oceangoing mosasaurs

and plesiosaurs, marine invertebrates like ammonites and belemnites, a variety of bivalves, and planktonic foraminifera. Marsupials, marine fishes, and
lizards barely made it through. Others—turtles, crocodiles, placental mammals—continued undisturbed.

This event, known as the Cretaceous-Tertiary (or K-T) extinction, was one
of the most dramatic of the mass kill-offs of geologic time, and it is one of the
best studied (chapter 8). Up until 1980, most of the serious ideas about the
K-T mass extinction revolved around changing climate caused by the loss of
midcontinental seas, shifts in continental positions, and changes in rates of
mountain building or volcanism. These kinds of events, operating very slowly—perhaps on the order of several million years—would have had an equally slow effect on Late Cretaceous life. One can imagine the slow elimination
of dinosaurs and their neighbors because of their inability to adapt to changing environmental conditions.

Iridium

But in 1980 the picture changed from slow and stately to catastrophic and
unpredictable. In that year a team of researchers from the University of California at Berkeley—led by paleontologist Walter Alvarez and his Nobel
Prize–winning father, Louis Alvarez—reported on a marine clay layer that
marks the K-T extinction at Gubbio, Italy. In this layer they found unusually high concentrations of the platinum metal iridium, up to thirty times more
than would be expected from normal influxes of cosmic dust (fig. 10.8). The
Alvarezes and their colleagues suggested that these high concentrations of
iridium must have come from space, where the metal is much more common. This enrichment of iridium in the Gubbio clay layer formed the basis
for what is now called the "asteroid hypothesis."

Let's look in some detail at the scenario of the asteroid hypothesis. According to the 1980 study, the asteroid that struck Earth at the end of the Cretaceous was probably about 5 to 10 kilometers (3 to 6 miles) in diameter, large
enough to produce the concentration of iridium at Gubbio. Traveling at
58,000 km an hour (36,000 miles an hour), it formed a fiery plume as it entered the atmosphere, punched a great hole in Earth's crust, and vaporized
all that it hit within seconds. As it drilled its way through the outer surface of
Earth, the asteroid ejected a dense, globe-circling cloud of molten rock, dust,
and debris. When it finally settled back to Earth, this dust gave us the iridi-
um-rich clay layer that marks the end of the Cretaceous.

But there is more to say about asteroids and extinction. Some have suggested that heat from the asteroid impact may have produced short-lived global warming. At ground zero, between 50 and 150 times the amount of solar energy that normally strikes Earth would have been released by the impact. Elsewhere temperatures may have risen as much as 30°C (86°F) for as long as a month, setting off global wildfires. Acid rain at least as bad as that seen in our modern cities may also have poisoned land and sea.

As the cloud of dust blotted out the sun's rays, global temperatures began to drop. This cooling may have been long term—from several months to perhaps years—compounding the other disasters brought on by asteroid impact. The successive cataclysms of heat, cold, acid rain, wildfire, and lack of sunlight first affected the plant realm, both terrestrial and marine. This led to starvation and death among plant-eating animals, then to the demise of their predators. Indeed, it is a surprise that anything survived the K-T asteroid impact.

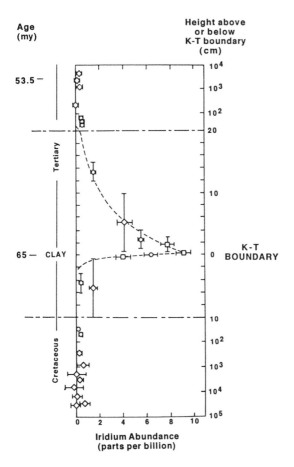

FIG. 10.8 Concentration of iridium at the Cretaceous-Tertiary boundary at Gubbio, Italy. At the boundary, the concentration is many times greater than is expected from the average dusting of iridium the planet gets through the ages (after Raup 1986).

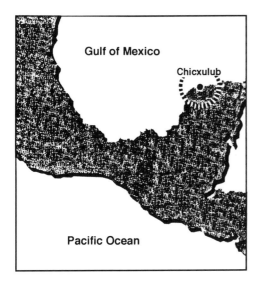

FIG. 10.9 The Chicxulub Crater along the margin of the Yucatán Peninsula of eastern Mexico, thought to represent ground zero for the dinosaur-killing asteroid (after Norman 1991).

Other Evidence

This fascinating killing scenario was highly speculative in 1980. But more than one hundred K-T boundary sites with elevated concentrations of iridium have since been found. In addition, other compounds that are much more common from extraterrestrial sources—such as osmium and certain kinds of amino acids—have also been identified from the K-T boundary.

Even more telling is the presence of shocked quartz and microtektites. Quartz grains, like the kind we find in beach sand, become compressed and deformed when subjected to immense and rapid pressures, a condition described as being "shocked." Microtektites are small, droplet-shaped blobs of silica-rich glass thrown up into the atmosphere in a molten state when a meteor strikes the earth. Both shocked quartz and microtektites are common in the K-T boundary, and they provide some of our strongest evidence of an asteroid impact 65 Ma.

But where is the all-powerful evidence of the impact itself? Where is the crater? Up until the mid-1980s, there were only guesses. Now it appears the "smoking gun" of the K-T boundary impact has been found, buried under meters of more recent sediment in the Yucatán Peninsula, beneath and offshore of the town of Chicxulub, Mexico (fig. 10.9). This 180 kilometer (112 mile) wide crater, dated at 65 million years ago, reveals the presence of shocked quartz and good chemical matches with microtektites from nearby sites.

Dinosaur Evidence from Montana

With so much evidence, the asteroid hypothesis is now widely accepted. But did it kill the dinosaurs? Some paleontologists have argued that dinosaurs had already been in decline for the last 10 million years or so of the Cretaceous, so there is no need to blame asteroids. Although it is true that dinosaur diversity appears to decrease about 10 million years before the K-T boundary, in some regions dramatically, it is also true that the number of species stays fairly constant from that time forward.

Paleontologist Peter Sheehan of the Milwaukee Public Museum and his colleagues have sampled the distribution of dinosaurs in the Hell Creek Formation of eastern Montana. They found that dinosaur diversity stays relatively constant from the bottom of the formation to the K-T boundary at its top. Thus there is no decrease in kinds of dinosaurs until the final moment marked by iridium, shocked quartz, and microtektites.

Eastern K-T Boundary

Here on the East Coast, we have a fair record of the mass extinction that marks the K-T boundary. Most recently studied by William B. Gallagher of the New Jersey State Museum (chapter 9), the Navesink and Hornerstown formations in New Jersey have an important tale to tell about the extinction. Although no iridium concentrations, microtektites, or shocked quartz have yet been identified at the New Jersey K-T boundary, clues can be drawn from the pattern of survival and extinction of Late Cretaceous and early Tertiary life.

As dinosaurs go, *Dryptosaurus, Coelosaurus,* and *"Hadrosaurus" minor* appear to be some of the last on the East Coast, accompanied by a few nondescript hadrosaurids and nodosaurids. But no remains of these dinosaurs have been found above the K-T boundary. Indeed, the only survivors of this once great terrestrial dynasty are those that took to the air: the diverse shorebirds from the Navesink and Hornerstown portion of the Upper Cretaceous.

For a better look at the K-T mass extinction along the East Coast, we need to examine the record of marine invertebrates. Although the extinction was relatively rapid, it appears to have been selective. For example, oysters and other mollusks dominated the ocean-bottom communities at the end of the Cretaceous. All began life as free-floating larvae that fed on microscopic plants and animals until they were ready to settle down and grow shells.

But in the earliest moments of the Tertiary things changed drastically. The fossil record tells us that mollusks no longer dominated. In their place were brachiopods, a variety of shellfish commonly known as lampshells, whose larvae did not depend on eating plankton. This pattern of selective extinction also applied to ammonites, the most common and diverse of invertebrate predators in the Mesozoic seas. Ammonites had larvae that fed on plankton, whereas their nautilid cousins laid large, self-sustaining eggs. At the K-T boundary, it was curtains for ammonites and clear sailing for nautilids.

Gallagher attributes this pattern of rapid, though selective, extinction to the population crash of planktonic organisms at the K-T boundary. This plankton crash is known worldwide and could have been triggered by an asteroid impact, perhaps through changes in the chemistry of the oceans. Although the East Coast missed out on the preservation of the geological signatures of an asteroid impact—iridium, microtektites, shocked quartz—we see the direct biological consequences of its profound disturbances.

East Coast Reprise

What can we expect for the future of East Coast dinosaur paleontology? We may learn more about the great extinctions at the end of the Triassic and the Cretaceous. With new discoveries, the dinosaur inhabitants of the great rift valleys of the Late Triassic and Early Jurassic may become more familiar. Perhaps we will get to know more about the trackmakers of this same time period through the discovery of skeletal fossils.

Even as the Lower Cretaceous exposures of Maryland and the District of Columbia are being paved over and built on, there are still remarkable finds being made. These discoveries increase our understanding of this important time in prehistory, when angiosperms were beginning their bloom, sauropods were in decline, and the diversification of ornithischians was under way.

But if the history of discovery is any measure, it is the Upper Cretaceous rocks of the East Coast and the dinosaurs they contain that may prove the most rewarding. During this time interval, dinosaur diversity was at its highest everywhere in the world. With diligence and a keen eye—as at Ellisdale, New Jersey (chapter 9)—it may be possible to sample more and more of all kinds of dinosaurs.

With continued studies of the skeletal remains and footprints already found in the East, we will better comprehend the relation of our dinosaurs to those found elsewhere: to the dinosaurs of the West, to those of Europe and Africa. And the discovery of more and better fossils of the dinosaurs that roamed the East Coast can only increase our understanding of worldwide dinosaur distribution and evolution.

11

E A S T   C O A S T   P A L E O P R I M E R

The fossil treasure of the East Coast reveals itself to those who know where and how to look. These patient seekers take satisfaction in the discovery of a single bone fragment or splintered tooth, sometimes from sites that vanish overnight. Because the dinosaur fossil record here is so sparse compared with collections from the North American West and rich fossil fields elsewhere in the world, every scrap of bone is potentially important. Scientists need as many clues as possible to piece together the great puzzle of the dinosaurs of the East Coast.

The Unique East Coast

Unlike many other dinosaur fossil localities in the world, East Coast sites are exposed less by nature than by human activities. Natural exposures of the fossil-bearing rock layers occur infrequently where streams or rivers or bodies of water like the Bay of Fundy in Nova Scotia have cut through the sediments. Sites such as Phoebus Landing in North Carolina and Ellisdale and Big Brook in New Jersey are examples of rare natural exposures along river or stream banks where dinosaur fossils have been found.

Abundant rainfall on the East Coast—five times the annual average in some western states—often hides rather than uncovers fossils. In the arid

West, broad areas of sedimentary rock are constantly eroded by wind, runoff from rain and snowstorms, and freezing and thawing. In the wet East, vegetation quickly claims most exposures and slows or halts the erosion that makes fossil discovery more likely. Many historic eastern dinosaur sites are now virtually impossible to locate, lost beneath an obscuring blanket of weeds, shrubs, trees, and leaves. In fact, fossils in many areas of the East Coast continue to be buried deeper as layer upon layer of dead organic matter accumulates above them in the ongoing natural process that creates topsoil.

From the very beginning of eastern dinosaur paleontology, most fossils have appeared as by-products of excavations: for tunnels, highway and railroad cuts, quarries, building and bridge foundations, water wells, and canals. When serious digging takes place in the right sediments, dinosaur bones and footprints are likely to be found. The challenge on the East Coast is to find and collect man-made exposures before they disappear, often under the vast carpet of concrete and asphalt that paves the megalopolis. Another obstacle to dinosaur hunters is the precipitous decline in quarrying industries that brought to light the rich fossil-bearing sediments of the Mesozoic.

But widespread development and large populations also benefit eastern dinosaur paleontology. Extensive excavation for buildings, roads, and parking lots means more exposures opened in the fossil sediments. And high population densities translate into a larger pool of potential fossil hunters.

Finders, Not Keepers

It's a maxim of paleontology that finding fossils is 90 percent luck and 10 percent skill, whereas collecting fossils is 90 percent skill and 10 percent luck. The suggested procedure for collectors who find a dinosaur fossil is to notify a museum, geological survey, natural history society, or fossil club. Specialists can then direct removal of the fossil to maximize its preservation and its benefit to the scientific community.

Before the Hunt

Serious hunters can increase their chances of finding dinosaur fossils on the East Coast by knowing where to begin the search. Fossils are discovered almost exclusively in sedimentary rocks, and dinosaur fossils are found only in sedimentary rocks of the correct age and conditions of deposition. Geologic maps are available from state geological surveys and the U.S. Geological Survey that locate Mesozoic rock formations in each individual state.

Collectors also need to study the historical record as a guide to their search. After two centuries of dinosaur hunting on the East Coast, some clear patterns have emerged:

- Dinosaur footprints are largely of Late Triassic and Early Jurassic age and have been found only in the Connecticut Valley, Nova Scotia, New York, New Jersey, Pennsylvania, Virginia, and Maryland.
- Early Cretaceous dinosaur skeletal remains have appeared only in the Arundel Clay of Maryland and Washington, D.C.
- Late Cretaceous dinosaur bones have come from marine greensand deposits in New Jersey, Delaware, and the Carolinas.
- No Middle-to-Late Jurassic dinosaur remains have been collected in eastern North America.

More Collecting Protocols

Collectors should observe a number of protocols in addition to notifying specialists about a fossil find. Most sites are on private property, and permission must be obtained from the owners before embarking on a fossil hunt. Managers of business properties such as quarries are particularly leery of amateur fossil seekers because of liability concerns and potential interference with work schedules. Quarries, road cuts, and railroad tunnels can be dangerous for inexperienced hunters, as can river sites that are reachable only by boat. Although sites owned by local, state, and federal governments are public property, permission for collecting must be obtained from the appropriate authorities.

The commonsense approach to fossil hunting is to ask permission, obey laws, show courtesy, and learn the ropes from knowledgeable collectors. Many fossil clubs sponsor field trips to provide supervised experience at local sites.

Collecting a Fossil

Some dinosaur fossils are found washed from their entombing rock, such as the stream-transported material at Ellisdale, New Jersey (chapter 9). But most fossils remain solidly embedded, and their safe collection is almost always a challenge. Before taking fossil discoveries from the field, paleontologists often compile detailed notes and draw diagrams of the dinosaur remains and their position relative to each other and surrounding rock layers. This information can be invaluable later to help assemble a skeleton or determine a dinosaur's age. In the case of footprints, the measurements taken may include compass direction, length of individual tracks, and stride distance between prints in trackways. The tracks of dinosaurs can reveal valuable clues to their physiology and behavior (chapter 2).

Other essential field documentation of fossils includes close-up photography. To provide size reference in the photographs, an object of known dimensions such as a coin or ruler is placed close to the fossil. Footprints can also be documented by "rubbing"—placing a sheet of white paper over the track and running the side of a soft pencil lead on the surface—or by "casting": applying liquid plaster or latex to the track to create a permanent record of the impression.

The classic method of removing dinosaur remains from rock in the field hasn't changed much since the mid-1800s. The paleontologist carefully exposes as much of the fossil as possible and sifts surrounding material for bone and tooth fragments. Small fossils are lifted from the rock, gently wrapped, and labeled for later study. If the bone is large and fragile, it may be treated with a hardener and then swathed in layers of gauzelike bandage soaked in plaster. When the plaster dries, the block can be safely dug out for transport back to the laboratory. Dinosaur footprints can often be removed as part of the large sandstone slabs they are imprinted on.

For Posterity's Sake

Not all fossils are worthy of special collecting care and permanent protection in a museum, but each dinosaur fossil found on the East Coast has the potential for great scientific interest. Thanks to the conscientious actions of the pioneers of eastern paleontology, some of the earliest dinosaur fossils are still safe and sound. Many are on exhibit for public viewing, and they are available to paleontologists for study and comparison with new discoveries. Among the largest collections of East Coast dinosaur fossils are those at the Smithsonian's U.S. National Museum of Natural History, the Academy of Natural Sciences of Philadelphia, the Peabody Museum of Natural History at Yale University, the New Jersey State Museum, the American Museum of Natural History, and the Pratt Museum of Natural History at Amherst College.

Fossil Preparation

In museum and university laboratories, skilled technicians known as preparators receive the fossils collected in the field and set about the laborious task of removing the remaining rock. Their arsenal includes small power tools, dental instruments, and sewing needles, often wielded while viewing the fossil under a microscope. Rock can be abraded away with miniature sandblasters, and certain kinds of rock can be removed by soaking the fossil in weak acid. Fossil pieces are then carefully glued together and treated with preservatives to prevent deterioration in storage. Casts, or exact replicas, of the fossil bones may be made out of plaster or fiberglass to permit a precious original to remain in safekeeping while the copy is put on display.

A modern development in laboratory paleontology involves the use of X rays and CT (computerized tomography) scans to look inside fossils that can't be separated from rock, such as fossil eggs and skulls filled with solid stone. CT scans have revealed tiny dinosaur embryos still inside their unhatched eggs.

As a footnote, fossil enthusiasts should consider volunteering in the paleontology departments of their nearest East Coast museums. Many of these institutions do not have the resources to maintain an adequate staff of preparators, and they would welcome the help.

1636 Harvard College founded.

1650 Archbishop James Ussher of Ireland sets date of creation as noon on October 23, 4004 B.C.

1676 Robert Plot, a clergyman at Oxford, describes dinosaur thighbone from England as part of a giant human.

1706 Cotton Mather proclaims mastodon fossils found near the Hudson River in 1705 to be remains of sinful giants drowned in Noah's flood.

1716 Yale College opens at New Haven, Connecticut.

1739 Charles Le Moyne of Canada discovers mastodon bones and teeth at Big Bone Lick along the Ohio River in Kentucky.

1743 American Philosophical Society organized in Philadelphia by Benjamin Franklin and James Logan.

1743 Mark Catesby, English naturalist, notes fossil shells and "elephant grinders" (mastodon teeth) in Virginia and Carolina, "imbedded a Great Depth in the Earth."

1763 English physician Richard Brookes names Plot's 1676 bone *Scrotum humanum* for its resemblance to giant human testicles.

1767 Benjamin Franklin observes that it "looks as if the earth had anciently been in another position, and the climates differently placed from what they are at present."

1768   English physician William Hunter reports that the Big Bone Lick remains belonged to a "giant elephant" [mastodon] that was carnivorous and "thank Heaven . . . is probably extinct."

1769   *Transactions of the American Philosophical Society* begins publication.

1780   American Academy of Arts and Sciences founded.

1781   Thomas Jefferson affirms the popular view that Mother Nature would not permit extinction of animal species.

1786   Peale Museum opens in Philadelphia as the first popular museum of natural history and art in North America.

1787   Large fossil foot bone (probably dinosaur) from New Jersey presented to the American Philosophical Society by Dr. Caspar Wistar and Timothy Matlack.

1799   Jefferson names *Megalonyx* (an extinct giant sloth) from remains found in a West Virginia cave.

1800   Baron Georges Cuvier, French anatomist, recognizes twenty-three species of extinct animals from their fossil bones.

1802   Mastodon displayed at Peale Museum is the first mounted fossil skeleton in North America.

1802   Pliny Moody plows up North America's first dinosaur tracks on his father's farm near South Hadley, Massachusetts.

1806   William Clark describes bone (possibly dinosaur) from Montana on his famous expedition with Meriwether Lewis.

1809   English geologist William Smith finds three large bone fragments in Sussex, later identified as the dinosaur *Iguanodon*.

1812   Academy of Natural Sciences of Philadelphia founded.

1817   The Lyceum of Natural History of New York organized.

1818   First bones from North America later confirmed as dinosaur found in a well in East Windsor, Connecticut.

1818   Remains of the world's first known dinosaur, *Megalosaurus,* found in England.

1818   *American Journal of Science* begins publication.

1820   Nathan Smith publishes on the East Windsor bones, later attributed to the prosauropod dinosaur *Anchisaurus colurus*.

1822   Mary Ann Mantell discovers fossil teeth of the world's second known dinosaur, *Iguanodon,* in Sussex, England.

1824   William Buckland names *Megalosaurus*.

1824   Construction begins on Chesapeake and Delaware Canal.

1825   Gideon Mantell, English physician and husband of Mary Ann Mantell, names *Iguanodon*.

1826   Jefferson dies.

1829   Samuel Morton publishes the first scientific report on Cretaceous fossils of the Chesapeake and Delaware Canal.

1835   Fossil footprints discovered in paving slabs at Greenfield, Massachusetts.

1835   Edward Hitchcock begins his investigation of Connecticut Valley fossil footprints.

1836   Hitchcock publishes the first scientific study of fossil footprints.

1838   Dinosaur bones discovered on the John E. Hopkins farm in Haddonfield, New Jersey.

1838   *Proceedings of the American Philosophical Society* begins publication.

1840   Association of American Geologists founded.

1841   *Proceedings of the Academy of Natural Sciences of Philadelphia* begins publication.

1842   British anatomist Sir Richard Owen names Dinosauria, a new suborder of reptiles, in the *Proceedings of the British Association for the Advancement of Science.*

1843   James Deane publishes on the Greenfield, Massachusetts, tracks.

1846   U.S. National Museum opens in Washington, D.C.

1848   American Association for the Advancement of Science founded.

1854   Owen and British wildlife artist Benjamin Waterhouse Hawkins collaborate on *Megalosaurus* and *Iguanodon* statues for London's Crystal Palace Park.

1855   Dinosaur skeleton (*Anchisaurus polyzelus*) discovered at Springfield, Massachusetts, during blasting for United States armory.

1856   Anatomist Joseph Leidy of Philadelphia officially describes the first dinosaurs in North America from teeth collected in western North America.

1858   Additional bones from Haddonfield site named by Leidy *Hadrosaurus foulkii,* world's first nearly complete dinosaur skeleton.

1858   Hitchcock publishes *Ichnology of New England.*

1858   Ebenezer Emmons publishes on North Carolina vertebrate fossils.

1859   Philip Tyson finds fossil teeth at Bladensburg, Maryland, first evidence of North American sauropod dinosaurs.

1859   Christopher Johnston studies the Tyson teeth, names the dinosaur *Astrodon.*

1859   Charles Darwin publishes *On the Origin of Species.*

1864   Hitchcock dies.

1865   Supplement to Hitchcock's *Ichnology* published.

1865   Leidy formally describes *Astrodon johnstoni* tooth.

1865   *Diplotomodon horrificus* tooth from Mullica Hill, New Jersey, described by Leidy.

1865   Philanthropist George Peabody endows Peabody Museum at Yale University and chair for his nephew, Othniel C. Marsh, as first professor of paleontology in North America.

1866   The carnivorous dinosaur *Laelaps* (now *Dryptosaurus*), discovered at Barnsboro, New Jersey, described by Edward D. Cope.

1868   *H. foulkii* becomes world's first mounted dinosaur skeleton, at Academy of Natural Sciences of Philadelphia.

1869   Cope names *Ornithotarsus immanis* (now *H. foulkii*), from bones discovered by Samuel Lockwood at Keyport, New Jersey.

1869   The duck-billed dinosaur *Hyspsibema* named by Cope from bones found near Faison, North Carolina.

1869   American Museum of Natural History in New York founded.

1870   Marsh describes the duckbill *Hadrosaurus minor* from New Jersey.

1877   Marsh renames *Laelaps* as *Dryptosaurus*.

1878   Cope describes the dinosaur *Thecodontosaurus gibbidens* (now *Galtonia gibbidens*) from teeth collected near Emigsville, Pennsylvania.

1884   Prosauropod dinosaur (*Ammosaurus major*) skeleton discovered in Wolcott Quarry at Manchester, Connecticut. Portion of skeleton lost when built into a bridge abutment.

1886   John Eyerman finds *Atreipus* footprints in quarry at Milford, New Jersey.

1887   John Bell Hatcher of Yale begins collecting Early Cretaceous dinosaurs for Marsh at Muirkirk, Maryland.

1888   Marsh names five new dinosaur species based on Hatcher collection from Maryland.

1891   Leidy dies.

1892   Owen dies.

1892   Marsh describes second prosauropod skeleton from Manchester quarry, now known as *Anchisaurus polyzelus*.

1892   Marsh describes third prosauropod skeleton from Manchester quarry, now thought to be a juvenile *Ammosaurus major*.

1894   Arthur Bibbins of Goucher College in Maryland begins collecting the Arundel Clay at Muirkirk.

1895   *Grallator* footprints discovered at Emmitsburg, Maryland, by James Mitchell.

1895   Carnegie Museum in Pittsburgh endowed by industrialist Andrew Carnegie.

1896   *Ornithotarsus immanis* footbones found by Lewis Woolman at Merchantville, New Jersey.

1897   Cope dies.

1897   Publication of *Natural History* magazine begun by American Museum of Natural History.

1898   Carnosaur bone found in Washington, D.C.

1899   Marsh dies.

1902   *Atreipus* tracks found at Fisher's Quarry near Graterford, Pennsylvania.

1904   Richard Swann Lull of Yale begins his research on the fossil footprints of the Connecticut Valley.

1904   Hatcher dies.

1905   Fossil collecting begins at Phoebus Landing, North Carolina.

1907   Bertram Boltwood of Yale University uses radioactive dating to put age of Earth at 1.6 billion years.

1910   Mignon Talbot of Mount Holyoke College discovers small dinosaur skeleton in South Hadley, Massachusetts.

1911   Lull revises Early Cretaceous dinosaur material from Maryland's Arundel Clay.

1911   Talbot describes her dinosaur skeleton as the theropod *Podokesaurus holyokensis*.

1915   Alfred Wegener introduces theory of continental drift.

1915   Lull publishes *Triassic Life of the Connecticut Valley.*

1916   Goucher College donates Bibbins collection to the U.S. National Museum.

1916   *Podokesaurus* skeleton destroyed in fire at Williston Hall, Mount Holyoke College.

1920   Dinosaur footprints found at Oak Hill estate in Virginia.

1921   Charles Gilmore of the U.S. National Museum revises Arundel dinosaur material.

1929   Tracks uncovered in Woodbridge, New Jersey, the only Cretaceous dinosaur footprints ever found in North America.

1933   Carlton Nash finds dinosaur tracks near the Pliny Moody locality in South Hadley, Massachusetts.

1936   Bibbins dies.

1937   *Atreipus* tracks found at Trostle Quarry near York Springs, Pennsylvania.

1939   Earl Poole locates dinosaur tracks at Schwenksville, Pennsylvania.

1942   *Astrodon* thighbone found in Washington, D.C.

1945   Gilmore dies.

1947   Discovery of *"Hadrosaurus" minor* remains at greensand pit in Sewell, New Jersey.

1948   Edwin Colbert describes *"H." minor* from Sewell.

1950   Talbot dies.

1952   Wilhelm Bock publishes "Triassic Reptilian Tracks and Trends of Locomotive Evolution."

1953   Lull publishes revision of *Triassic Life.*

1957   Lull dies.

1964   John Ostrom of Yale discovers *Deinonychus* in Montana, marking the advent of modern dinosaur research.

A P P E N D I X T W O
D I N O S A U R D I R E C T O R Y

Museums

The following institutions display or archive East Coast dinosaur fossils and can provide information about local fossil collecting:

Connecticut

Dinosaur State Park
West Street
Rocky Hill, CT 06067
(203) 529-8423

Peabody Museum of Natural History
Yale University
170 Whitney Avenue
New Haven, CT 06511
(203) 432-5050

Delaware

See New Jersey: New Jersey State
  Museum

Maryland

See Washington, D.C.: National
  Museum of Natural History

Massachusetts

Agassiz Museum of Comparative
  Zoology
Oxford Street
Cambridge, MA 02138
(617) 495-3045

Pratt Museum of Natural History
Amherst College
Amherst, MA 01002
(413) 542-2165

Springfield Science Museum
236 State Street
Springfield, MA 01103
(413) 733-1194

Mount Holyoke College
35 Woodbridge Street
South Hadley, MA 01015
(413) 538-2085

New Jersey

New Jersey State Museum
205 West State Street
Trenton, NJ 08608
(609) 292-6308

Museum of Natural History
Princeton University
Washington Street
Princeton, NJ 08544
(609) 258-1322

Rutgers Geology Museum
Rutgers University
College Avenue
New Brunswick, NJ 08903
(908) 932-7243

New York

American Museum of Natural History
Central Park West at 79th Street
New York, NY 10024
(212) 769-5100

Buffalo Museum of Science
1020 Humboldt Parkway
Buffalo, NY 14211
(716) 896-5200

New York State Museum
Cultural Education Center
Empire State Plaza
Albany, NY 12230
(518) 474-5877

North Carolina

Cliffs of the Neuse State Park
345-A Park Entrance Road
Seven Springs, NC 28578
(919) 778-6234

North Carolina State Museum
102 North Salisbury Street
Raleigh, NC 27626
(919) 733-7450

Nova Scotia

Fundy Geological Museum
6 Two Islands Road
P.O. Box 640
Parrsboro, NS BOM 1S0
(902) 254-3814

Nova Scotia Museum
1747 Summer Street
Halifax, NS B3H 3A6
(902) 429-4610

Pennsylvania

Academy of Natural Sciences
19th Street and the Parkway
Philadelphia, PA 19103
(215) 299-1000

Carnegie Museum of Natural History
440 Forbes Avenue
Pittsburgh, PA 15213
(412) 622-3131

The State Museum of Pennsylvania
Third and North Streets
Harrisburg, PA 17120
(717) 787-4980

South Carolina

South Carolina State Museum
301 Gervais Street
P.O. Box 100107
Columbia, SC 29202
(803) 737-4921

Virginia

Virginia Museum of Natural History
1001 Douglas Avenue
Martinsville, VA 24122
(703) 666-8600

Museum of the Geological Sciences
Virginia Polytechnic Institute and
   State University
Department of Geological Sciences
2062 Derring Hall
Blacksburg, VA 24061
(703) 231-6521

Washington, D.C.

U.S. National Museum of Natural
   History
Smithsonian Institution
Tenth Street and Constitution Avenue
   NW
Washington, DC 20560
(202) 357-2700

Geological Surveys

   Geological maps of the East Coast and individual states can be ordered from
the following agencies:

U.S. Geological Survey
Earth Science Information Center
12201 National Center
Reston, VA 22092
(800) USA-MAPS

Connecticut Geological and Natural
   History Survey
Map and Publication Sales
165 Capitol Avenue, Room 555
Hartford, CT 06106
(203) 566-7719

Delaware Geological Survey
University of Delaware
Newark, DE 19716
(302) 831-2833

Maryland Geological Survey
2300 St. Paul Street
Baltimore, MD 21218
(410) 554-5505

Massachusetts Office of
  Environmental Affairs
100 Cambridge Street
Boston, MA 02202
(617) 727-9800

New Jersey Geological Survey
CN-427
Trenton, NJ 08625
(609) 292-1185

New York State Museum
Publication Sales
3140 Cultural Education Center
Albany, NY 12230
(518) 474-3505

North Carolina Geological Survey
P.O. Box 27687
Raleigh, NC 27611
(919) 733-2423

Pennsylvania Topographic and
  Geologic Survey
P.O. Box 8453
Harrisburg, PA 17105
(717) 787-2169

South Carolina Geological Survey
5 Geology Road
Columbia, SC 29210
(803) 737-9440

Virginia Division of Mineral
  Resources
P.O. Box 3667
Charlottesville, VA 22903
(804) 293-5121

A P P E N D I X   T H R E E
D I N O S A U R   S I T E   M A P S

Map 1 Connecticut Valley: Massachusetts and Connecticut (Upper Triassic, Lower Jurassic).

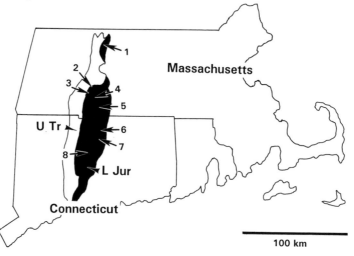

1. Turners Falls, Massachusetts
2. Pliny Moody's farm, Massachusetts
3. Mt. Tom, Massachusetts
4. South Hadley, Massachusetts

5. Springfield, Massachusetts
6. East Windsor, Connecticut
7. Manchester, Connecticut
8. Rocky Hill, Connecticut

Map 2 Virginia (Upper Triassic, Lower Jurassic, Lower Cretaceous).

1. Leaksville Junction, Virginia
2. Culpeper (Stevensburg), Virginia
3. Manassas, Virginia
4. Aldie (Oak Hill), Virginia

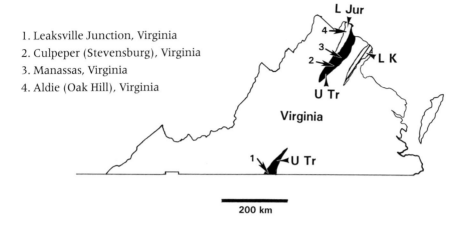

Map 3 Nova Scotia (Upper Triassic, Lower Jurassic).

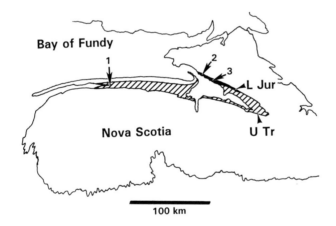

1. Digby, Nova Scotia
2. Parrsboro (Wasson Bluff), Nova Scotia
3. Paddy Island, Nova Scotia

Map 4 Pennsylvania, New Jersey, New York (Upper Triassic, Lower Jurassic, Upper Cretaceous).

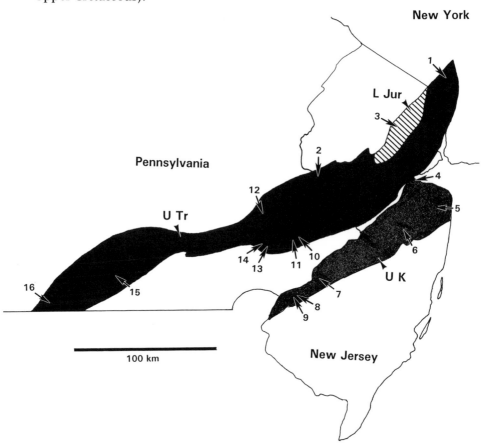

1. Nyack Beach State Park, Haverstraw, New York
2. Milford, New Jersey
3. Roseland (Riker Hill), New Jersey
4. Woodbridge, New Jersey
5. Big Brook, New Jersey
6. Ellisdale, New Jersey
7. Haddonfield, New Jersey
8. Barnsboro, New Jersey
9. Sewell, New Jersey
10. Gwynedd, Pennsylvania
11. Limerick, Pennsylvania
12. Reading, Pennsylvania
13. Graterford, Pennsylvania
14. Schwenksville, Pennsylvania
15. Emmigsville, Pennsylvania
16. Gettysburg, Pennsylvania

Map 5 Maryland, Delaware, Washington, D.C. (Upper Triassic, Lower Cretaceous, Upper Cretaceous).

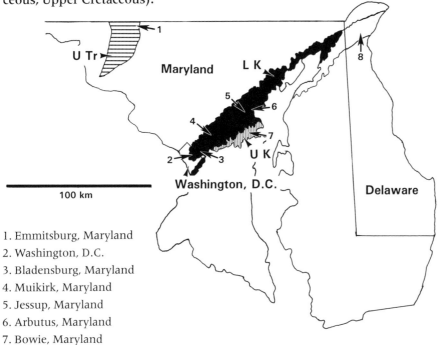

1. Emmitsburg, Maryland
2. Washington, D.C.
3. Bladensburg, Maryland
4. Muikirk, Maryland
5. Jessup, Maryland
6. Arbutus, Maryland
7. Bowie, Maryland
8. Chesapeake and Delaware Canal, St. George, Delaware

Map 6 North Carolina, South Carolina (Upper Triassic, Upper Cretaceous).

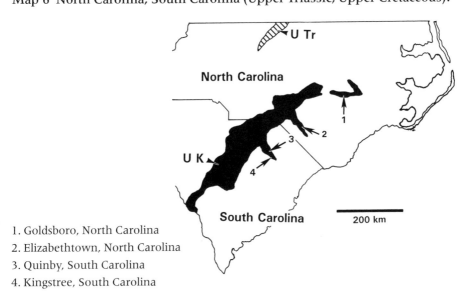

1. Goldsboro, North Carolina
2. Elizabethtown, North Carolina
3. Quinby, South Carolina
4. Kingstree, South Carolina

REFERENCES

Alexander, R. McN. 1989. *Dynamics of dinosaurs and other extinct giants.* New York: Columbia University Press.

Alexander, T. 1975a. A revolution called plate tectonics has given a whole new understanding of the configuration of the earth—and how and why it came about. *Smithsonian* 5:30–40.

———. 1975b. Plate tectonics has a lot more to tell us about the earth as it is—and as it will be. *Smithsonian* 5:38–47.

Allaby, M., and J. Lovelock. 1983. *The great extinction.* New York: Doubleday.

Allyn, H. M. 1950. A tribute to Mignon Talbot. *Mount Holyoke Alumnae Quarterly* 34:112.

Alvarez, L. W., W. Alvarez, F. Asaro, and H. V. Michel. 1980. Extraterrestrial cause for the Cretaceous-Tertiary extinction. *Science* 208:1095–1108.

Alvarez, W., and F. Asaro. 1990. An extraterrestrial impact. *Scientific American* 263:78–84.

Annan, R. 1793. Account of a large animal found near Hudson's River. *American Academy of Arts and Sciences Memoirs* 2:160–164.

[Anonymous.] 1833. Museum of Gideon Mantell, Esq., of Lewes, in Sussex, England. *American Journal of Science* 23:162–179.

———. 1885. The relationships between dinosaurs and birds. *Science* 6:295.

———. 1910. Department notes: Geology. *Mount Holyoke* 20 (December): 285–286.

———. 1911. Department notes: Geology. *Mount Holyoke* 20 (February): 402.

———. 1912. Department notes: Geology. *Mount Holyoke* 21 (May): 483.

———. 1916. List of cycads and vertebrate fossils deposited in the National Museum by Goucher College. Goucher College Archives, Towson, Md.

———. 1922. The first mounted skeleton of a dinosaur, *Hadrosaurus. Natural History* 22:382–393.

———. 1923. Dinosaur remains near New York. *Natural History* 23:98.

———. 1934. Rare dinosaur tracks found in Pennsylvania. *Science News Letter,* July 28, 57.

———. 1936. Obituary for Arthur Barneveld Bibbins. *Goucher Alumnae Quarterly* 15:17.

———. 1937a. Dinosaur tracks found near Gettysburg site. *Science News Letter,* December 18, 392.

———. 1937b. Quarry gives up dinosaur footprint after millions of years. *Pennsylvania Department of Internal Affairs Monthly Bulletin* 4:115–123.

———. 1940. *Guide to the Old Line State.* Washington, D.C.: Works Progress Administration.

———. 1942a. Dinosaur find at Washington, D.C. *Nature* 150:54.

———. 1942b. Thighbone of dinosaur found in Washington, D.C. *Science News Letter,* May 23, 328.

———. 1945. Dinosaur footprints in Pennsylvania. *Rocks and Minerals* 20:423.

———. 1957. Dinosaur's thigh bone found in marl pit. *Science News Letter,* July 6, 9.

———. 1958. Dollars from dinosaurs: Dinosaur tracks. *Newsweek,* August 18, 68.

———. 1967a. Connecticut: A vast number of dinosaur footprints were unearthed near the town of Rocky Hill. *Rocks and Minerals* 42:45.

———. 1967b. *Inland waterway, Delaware River to Chesapeake Bay, historic Chesapeake and Delaware Canal.* Washington, D.C.: U.S. Army Corps of Engineers, U.S. Government Printing Office.

———. 1969a. The missing *Ammosaurus:* Discovered in a stony quarry near Manchester, Connecticut. *Time,* November 7, 53.

———. 1969b. Brownstone dinosaur. *Scientific American* 221 (October): 50.

———. 1970. Dinosaurs finally win one: P. Olsen and T. Lessa efforts to save fossil-bearing acres for an educational park. *Life,* December 11, 73–74.

———. 1972. *Helpful hints for fossil collecting in the Chesapeake and Delaware Canal.* Washington, D.C.: U.S. Army Corps of Engineers, U.S. Government Printing Office.

———. 1974. *The Chesapeake and Delaware Canal.* Washington, D.C.: U.S. Army Corps of Engineers, U.S. Government Printing Office.

———. 1979. *The Chesapeake and Delaware Canal.* Washington, D.C.: U.S. Army Corps of Engineers, U.S. Government Printing Office.

———. 1980a. Tracks of swimming dinosaurs challenge old theories. *New Scientist* 85:1010.

———. 1980b. Carnivorous dinosaurs in the swim. *Science News,* March 22, 181.

———. 1980c. Dinosaur tracks make a comeback (C. S. Nash marketing operation in Massachusetts). *Nation's Business,* October, 86.

———. 1981. *Dinosaur footprints found at Philadelphia Electric Company's Limerick Generating Station.* Pamphlet. Philadelphia: Philadelphia Electric Company.

———. 1986a. The littlest dinosaur (Nova Scotia). *Discover* 7:13.

———. 1986b. Finding fossilized clues to the past (Nova Scotia). *U.S. News and World Report,* February 17, 66.

———. 1989. Dinosaur tracks found, on exhibit in Virginia (Culpeper). *Earth Science* 42(fall):8.

———. 1990. Tiny dinosaurs shed light on mass extinctions. *Earth Science* 43(summer): 11–12.

R E F E R E N C E S

————. 1991a. Ancient dinosaur prints found at Virginia construction site. *Montgomery Journal,* October 3, 4A.

————. 1994. *Dr. Earl L. Poole, 1891–1972: Artist, illustrator, gentleman, scholar.* Commemorative pamphlet. Reading, Pa.: Historical Society of Berks County.

Atwater, C. 1820. On some ancient human bones, etc., with a notice of the bones of the mastodon or mammoth, and of various shells found in Ohio and the West. *American Journal of Science* 2:242–246.

Bain, G. W., and B. W. Harvey. 1977. Field guide to the geology of the Durham Triassic Basin. Paper presented at Carolina Geological Society fortieth anniversary meeting. North Carolina Department of Natural Resources and Community Development, Raleigh.

Bain, G. W., and H. Meyerhoff. 1976. *The flow of time in the Connecticut valley.* Springfield, Mass.: Connecticut Valley History Museum.

Baird, D. 1954. *Cheirotherium lulli,* a pseudosuchian reptile from New Jersey. *Harvard Museum of Comparative Zoology Bulletin* 111:165–192.

————. 1957. Triassic reptile footprint faunules from Milford, New Jersey. *Harvard Museum of Comparative Zoology Bulletin* 117:449–520.

————. 1966. What's new in the Cretaceous? *Delaware Valley Earth Sciences Society News Bulletin* 3:81–85.

————. 1967. Age of fossil birds from the greensands of New Jersey. *Auk* 84:260–262.

————. 1984. Lower Jurassic dinosaur footprints in Nova Scotia. *Ichnology Newsletter* 14:2.

————. 1986a. Upper Cretaceous reptiles from the Severn Formation of Maryland. *Mosasaur* 3:63–85.

————. 1986b. Some Upper Triassic reptiles, footprints, and an amphibian from New Jersey. *Mosasaur* 3:125–153.

————. 1989. Medial Cretaceous carnivorous dinosaur and footprints from New Jersey. *Mosasaur* 4:53–63.

————. 1993. The Trostle Quarry footprints and their makers. In *Guidebook for the twelfth annual field trip,* 36–37. Harrisburg, Pa.: Harrisburg Area Geological Society.

Baird, D., and G. R. Case. 1966. Rare marine reptiles from the Cretaceous of New Jersey. *Journal of Paleontology* 40:1211–1215.

Baird, D., and P. M. Galton. 1981. Pterosaur bones from the Upper Cretaceous of Delaware. *Journal of Vertebrate Paleontology* 1, no. 1:67–71.

Baird, D., and J. R. Horner. 1977. A fresh look at the dinosaurs of New Jersey and Delaware. *New Jersey Academy of Science Bulletin* 22:50.

————. 1979. Cretaceous dinosaurs of North Carolina. *Brimleyana,* no. 2:1–28.

Baird, D., and P. E. Olsen. 1983. Late Triassic herpetofauna from the Wolfeville Formation of the Minas Basin of Nova Scotia. Geological Society of America, annual meeting, Northeastern Section. *Abstracts with Programs* 15:122.

Baird, D., and W. F. Take. 1959. Triassic reptiles from Nova Scotia. *Geological Society of America Bulletin* 70:1565–1566.

Bakker, R. T. 1975. Dinosaur renaissance. *Scientific American* 232:58–72.

————. 1986. *The dinosaur heresies.* New York: William Morrow.

Bakker, R. T., and P. M. Galton. 1974. Dinosaur monophyly and a new class of vertebrates. *Nature* 248:168–172.

Barbour, E. H. 1890. Notes on the palaeontological laboratory of the United States Geological Survey under Professor Marsh. *American Naturalist* 24:388–400.

Barrell, J. 1915. *Central Connecticut in the geologic past.* Bulletin 23. Hartford: Connecticut Geological and Natural History Survey.

Barton, B. S. 1804. Letter from James Wright. *Philadelphia Medical and Physical Journal* 1:154–159.

Bedini, S. A. 1985. *Thomas Jefferson and American vertebrate paleontology.* Publication 61. Charlottesville: Virginia Division of Mineral Resources.

Bell, W. J., Jr. 1949. A box of old bones: A note on the identification of the mastodon, 1766–1806. *American Philosophical Society Proceedings* 93:169–177.

Benton, M. J. 1983. Dinosaur success in the Triassic: A noncompetitive ecological model. *Quarterly Review of Biology* 58:29–55.

———. 1984. Dinosaurs' lucky break. *Natural History* 93:54–59.

———. 1991. What really happened in the Late Triassic? *Historical Biology* 5:263–278.

Berry, E. W. 1911. Correlation of the Potomac Formations. In *Maryland Geological Survey Lower Cretaceous,* 153–172. Baltimore: Johns Hopkins Press.

———. 1916. Systematic paleontology, Upper Cretaceous: Vertebrata. In *Maryland Geological Survey Upper Cretaceous,* 347–360. Baltimore: Johns Hopkins Press.

Bibbins, A. B. 1895. Notes on the paleontology of the Potomac Formation. *Johns Hopkins University Circular* 15, no. 121:17–20.

Bock, W. 1952. Triassic reptilian tracks and trends of locomotive evolution, with remarks on correlation. *Journal of Paleontology* 26:395–433.

Bower, B. 1986. Nova Scotia fossils illuminate 200 million-year-old changes. *Science News,* February 8, 86.

Brenner, G. J. 1963. *The spores and pollen of the Potomac Group of Maryland.* Bulletin 27. Baltimore: Maryland Department of Geology, Mines, and Water Resources.

Brett, C. E., and W. H. Wheeler. 1961. A biostratigraphic evaluation of the Snow Hill Member, Upper Cretaceous of North Carolina. *Southeastern Geology* 3:49–132.

Brett-Surman, M. K., and C. S. Martin. 1992. A preliminary report on new dinosaur material from the Arundel Clay (Lower Cretaceous, Late Aptian–Early Albian) of Maryland. Manuscript.

Brown, R. W. 1943. Jefferson's contribution to paleontology. *Journal of the Washington Academy of Sciences* 33:257–259.

Buckland, W. 1824. Notice on the *Megalosaurus* or great fossil lizard of Stonesfield. *Geological Society of London Transactions* 21:390–397.

———. 1841. *The Bridgewater treatises on the power, wisdom, and goodness of God as manifested in the Creation: Geology and mineralogy considered with reference to natural theology.* 2 vols. Treatise 6. Philadelphia: Lea and Blanchard.

Bukowski, F. 1979. Prehistoric residents of Essex County, New Jersey: Walter T. Kidde Dinosaur Park. *Earth Science* 32(summer): 111–112.

———. 1980. Cretaceous fossils from New Jersey and Delaware. *Earth Science* 33:55–60.

Burns, J. 1991. *Fossil collecting in the Mid-Atlantic states.* Baltimore: Johns Hopkins University Press.

Byrnes, J. B. 1972. The bedrock geology of Dinosaur State Park, Rocky Hill, Connecticut. Master's thesis, Department of Geology, University of Connecticut, Storrs.

R E F E R E N C E S

Carpenter, K. 1980. Washing and screening for small fossils. *Earth Science* 33(winter): 26–27.

Carr, M. S. 1950. *The District of Columbia: Its rocks and their geological history.* Bulletin 967. Washington, D.C.: U.S. Geological Survey.

Carrier, D. R. 1987. The evolution of locomotor stamina in tetrapods: Circumventing a mechanical constraint. *Paleobiology* 13:326–341.

Carroll, R. L. 1988. *Vertebrate paleontology and evolution.* New York: W. H. Freeman.

Carroll, R. L., E. S. Belt, D. I. Dineley, D. Baird, and D. C. McGregor. 1972. *Vertebrate paleontology of eastern Canada.* Guidebook of Field Excursions A59. Montreal: International Geological Congress.

Carter, C. W. 1937. The Upper Cretaceous deposits of the Chesapeake and Delaware Canal. *Maryland Geological Survey* 13, part 6:237–281.

Casler, K. 1989. Dinosaur tracks stolen. *Allentown Morning Call,* June 21, A1.

Charig, A. J. 1979. *A new look at the dinosaurs.* New York: Mayflower Books.

Charig, A. J., and B. Horsfield. 1975. *Before the ark.* London: British Broadcasting Corporation.

Chester, F. D. 1884. Preliminary notes on the geology of Delaware: Laurentian, Paleozoic, and Cretaceous areas. *Academy of Natural Sciences of Philadelphia Proceedings* 84:237–259.

Chinsamy, A. 1990. Physiological implications of the bone histology of *Syntarsus rhodesiensis* (Saurischia: Theropoda). *Palaeontologia Africana* 27:77–82.

Chure, D. J., and J. F. McIntosh. 1989. *A bibliography of the Dinosauria (exclusive of the Aves) 1677–1986.* Paleontology Series 1. Grand Junction: Museum of Western Colorado.

Clark, J. M. 1903. Mastodons of New York: A list of discoveries of their remains, 1705–1902. *New York Geological Survey Bulletin* 69:921–933.

Clark, W. B. 1916. The Upper Cretaceous deposits of Maryland. In *Maryland Geological Survey Upper Cretaceous,* 23–110. Baltimore: Johns Hopkins Press.

Clark, W. B., R. M. Bagg, and G. G. Shattuck. 1897. Upper Cretaceous formations of New Jersey, Delaware, and Maryland. *Geological Society of America Bulletin* 8:315–358.

Clark, W. B., E. W. Berry, and J. A. Gardner. 1916. Correlation of the Upper Cretaceous formations. In *Maryland Geological Survey Upper Cretaceous,* 315–342. Baltimore: Johns Hopkins Press.

Clark, W. B., and A. B. Bibbins. 1897. Stratigraphy of the Potomac Group in Maryland. *Journal of Geology* 5:479–506.

Clark, W. B., A. B. Bibbins, and E. W. Berry. 1911. The Lower Cretaceous deposits of Maryland. In *Maryland Geological Survey Lower Cretaceous,* 23–98. Baltimore: Johns Hopkins Press.

Cohen, D., and S. Cohen. 1992. *Where to find dinosaurs today.* New York: Cobblehill Books.

Colbert, E. H. 1946. *Hypsognathus,* a Triassic reptile from New Jersey. *American Museum of Natural History Bulletin* 86:225–274.

———. 1948. A hadrosaurian dinosaur from New Jersey. *Academy of Natural Sciences of Philadelphia Proceedings* 100:23–27.

———. 1960. A New Triassic procolophonid from Pennsylvania. *American Museum Novitates,* no. 2022:1–19.

———. 1961a. *The world of dinosaurs.* New York: Home Library Press.

———. 1961b. *Dinosaurs, their discovery and their world.* New York: E. P. Dutton.

————. 1963. *Fossils of the Connecticut Valley: The age of dinosaurs begins.* Bulletin 96. Hartford: Connecticut Geological and Natural History Survey.

————. 1964a. The Triassic dinosaur genera *Podokesaurus* and *Coelophysis. American Museum Novitates,* no. 2168:1–12.

————. 1964b. The fossils of New Jersey. *Nature News* 19, no. 1:17–27.

————. 1965. A phytosaur from North Bergen, New Jersey. *American Museum Novitates,* no. 2230:1–25.

————. 1966. A gliding reptile from the Triassic of New Jersey. *American Museum Novitates,* no. 2246;1–23.

————. 1968. *Men and dinosaurs: The search in field and laboratory.* New York: E. P. Dutton.

————. 1970a. The Triassic gliding reptile *Icarosaurus. American Museum of Natural History Bulletin* 143:89–142.

————. 1970b. *Fossils of the Connecticut Valley: The age of dinosaurs begins.* Rev. ed. Bulletin 96. Hartford: Connecticut Geological and Natural History Survey.

————. 1989. The Triassic dinosaur *Coelophysis. Museum of Northern Arizona Bulletin* 57:1–174.

Colbert, E. H., and D. Baird. 1958. Coelurosaur bone casts from the Connecticut Valley Triassic. *American Museum Novitates,* no. 1901:1–11.

Colton, H. S. 1909. Peale's museum. *Popular Science Monthly* 75:221–238.

Cooke, C. W. 1952. Sedimentary deposits of Prince Georges County and the District of Columbia. *Maryland Department of Geology, Mines, and Water Resources Bulletin* 10:1–53.

Cooke, C. W., and L. W. Stephenson. 1928. The Eocene age of the supposed late Upper Cretaceous greensand marls of New Jersey. *Journal of Geology* 36:139–148.

Coombs, W. P., Jr. 1980. Swimming ability of carnivorous dinosaurs. *Science* 207:1198–1200.

————. 1990. Behavior patterns of dinosaurs. In *The Dinosauria,* ed. D. B. Weishampel, P. Dodson, and H. Osmólska, 32–42. Berkeley: University of California Press.

Cope, E. D. 1866a. Discovery of a gigantic dinosaur in the Cretaceous of New Jersey. *Academy of Natural Sciences of Philadelphia Proceedings,* 275–279.

————. 1866b. [Remarks on *Laelaps*]. *Academy of Natural Sciences of Philadelphia Proceedings,* 316–317.

————. 1866c. Observations on extinct vertebrates of the Mesozoic red sandstone from Phoenixville, Chester County, Pennsylvania. *Academy of Natural Sciences of Philadelphia Proceedings,* 249–250.

————. 1867. The fossil reptiles of New Jersey. *American Naturalist* 1:23–30.

————. 1868a. Remarks on the freshwater origin and relations of certain sands and clays in New Jersey, Maryland, and Virginia. *Academy of Natural Sciences of Philadelphia Proceedings,* 157–158.

————. 1868b. Synopsis of the extinct Batrachia of North America. *Academy of Natural Sciences of Philadelphia Proceedings,* 208–221.

————. 1868c. On some Cretaceous reptilia. *Academy of Natural Sciences of Philadelphia Proceedings,* 233–242.

————. 1868d. On the genus *Laelaps. American Journal of Science* 46:415–417.

————. 1869a. The fossil reptiles of New Jersey. *American Naturalist* 3:84–91.

————. 1869b. Observations on a very large dinosaur in the collection of Dr. Samuel Lock-

wood, of Keyport, Monmouth County, New Jersey, discovered in the Lower Cretaceous clays on the shores of Raritan Bay. *Academy of Natural Sciences of Philadelphia Proceedings,* 123.

———. 1869c. [Remarks on *Holops brevispinus, Ornithotarsus immanis,* and *Macrosaurus
proriger*]. *Academy of Natural Sciences of Philadelphia Proceedings,* 123.

———. 1869d. Observations on the remains of reptiles from Sampson County, North Carolina, of Cretaceous age. *Academy of Natural Sciences of Philadelphia Proceedings,* 191–192.

———. 1869e. [Remarks on *Eschrichtius polyporus, Hypsibema crassicauda, Hadrosaurus tri-
pos,* and *Polydectes biturgidus*]. *Academy of Natural Sciences of Philadelphia Proceedings,*
191–192.

———. 1869f. Synopsis of the extinct Batrachia, Reptilia, and Aves of North America.
American Philosophical Society Transactions, N.S., 14:1–252.

———. 1870a. On the *Megadactylus polyzelus* of Hitchcock. *American Journal of Science*
44:390–392.

———. 1870b. Reptilia of the Triassic formations of the United States. *American Naturalist* 4:562–563.

———. 1871a. Supplement to the "Synopsis of the extinct Batrachia and Reptilia of North
America." *American Philosophical Society Proceedings* 12:41–52.

———. 1871b. Observations on *Hadrosaurus cavatus,* indicated by remains derived from
the upper green sand bed of the Upper Cretaceous of New Jersey. *American Philosophical Society Proceedings,* 50–52.

———. 1871c. Observations on the distribution of certain extinct Vertebrata in North Carolina. *American Philosophical Society Proceedings* 41:210–216.

———. 1878a. On some saurians found in the Triassic of Pennsylvania by C. M. Wheatley. *American Philosophical Society Proceedings* 17:231–232.

———. 1878b. The affinities of the Dinosauria. *American Naturalist* 12:57–58.

———. 1878c. Triassic saurians from Pennsylvania. *American Naturalist* 12:58.

———. 1889. On a new genus of Triassic Dinosauria. *American Naturalist* 23:626.

Cornet, B. 1977. The palynostratigraphy and age of the Newark Supergroup. Ph.D. diss.
Department of Geosciences, Pennsylvania State University.

———. 1989. Richmond Basin lithostratigraphy and paleoenvironments. In *Field guide to
the tectonics, stratigraphy, sedimentology, and paleontology of the Newark Supergroup, eastern
North America,* ed. P. E. Olsen and P. J. W. Gore, 47–52. Field Trip Guidebook T351.
Washington, D.C.: International Geological Congress.

Cornet, B., and P. E. Olsen. 1990. *Early to middle Carnian (Triassic) flora and fauna of the Richmond and Taylorsville basins, Virginia and Maryland, USA.* Guidebook no. 1. Martinsville:
Virginia Museum of Natural History.

Cornet, B., and A. Traverse. 1975. Palynological contribution to the chronology and
stratigraphy of the Hartford Basin in Connecticut and Massachusetts. *Geoscience and
Man* 11:1–33.

Cornet, B., A. Traverse, and N. G. McDonald. 1973. Fossil spores, pollen, and fish indicate
Early Jurassic age for part of the Newark Group. *Science* 182:1243–1247.

Crater, N. 1983. 220-million year old fossils are discovered. *Mercury* (Pottstown, Pa.), June
8, 1, 5.

Curran, H. A. 1985. The trace fossil assemblage of a Cretaceous nearshore environment:
Englishtown Formation of Delaware, USA. In *Biogenic structures,* ed. H. A. Curran,

261–276. Special publication 35. Tulsa, Okla.: Society of Economic Paleontologists and Mineralogists.

Cushman, J. A. 1904. A new foot-print from the Connecticut Valley. *American Geologist* 33:154–156.

Dana, J. D. 1896. *Manual of geology: Treating of the principles of the science with special reference to American geological history, etc.* 4th ed. New York: American Book Co.

Darton, N. H. 1891. Mesozoic and Cenozoic formations of eastern Virginia and Maryland. *Geological Society of America Bulletin* 2:431–450.

Darwin, C. R. 1859. *On the origin of species by means of natural selection, or The preservation of favoured races in the struggle for life; and The descent of man and selection in relation to sex.* London: John Murray.

Deane, J. 1843. On the fossil footmarks of Turner's Falls, Massachusetts. *American Journal of Science* 46: 73–77.

———. 1849. Illustration of fossil footprints of the valley of the Connecticut. *American Academy of Arts and Sciences Transactions 4:139–221.*

———. 1861. *Ichnographs from the sandstone of the Connecticut River.* Boston: Little, Brown.

deBoer, J. 1968. Late Triassic volcanism in the Connecticut valley and related structure. In *Guidebook for fieldtrips in Connecticut*, ed. P. M. Orville, trip C-5, sec. C-5, 1–12. New England Intercollegiate Geology Conference, annual meeting, Yale University. Hartford: Connecticut Geological and Natural History Survey.

Decina, L. 1985. Third reported occurrence of a hadrosaurian tooth found at the Chesapeake and Delaware Canal. *Delaware Valley Paleontological Society Newsletter* 7:2.

DeKay, J. E. 1830. On the remains of extinct reptiles of the genera *Mosasaurus* and *Geosaurus* found in the Secondary Formation of New Jersey; and on the occurrence of the substance recently named coprolite by Dr. Buckland, in the same locality. *Annals of the Lyceum Natural History of New York* 3:134–141.

Denton, R. K., Jr. 1990. A revision of the theropod *Dryptosaurus (Laelaps) aquilunguis* (Cope 1869). *Journal of Vertebrate Paleontology* 10, no. 3, suppl.: 20A.

Denton, R. K., Jr., and W. B. Gallagher. 1989. Dinosaurs of the Ellisdale site, Late Cretaceous (Campanian) of New Jersey. *Journal of Vertebrate Paleontology* 9, no. 3, suppl.: 18A.

Denton, R. K., Jr., R. C. O'Neill, B. S. Grandstaff, and D. C. Parris. 1993. The Ellisdale dinosaur site: An overview of its discovery, geology, and paleontology. Manuscript.

Denton, R. K., Jr., and D. C. Parris. 1985. A vertebrate fauna from the Marshalltown Formation (Late Cretaceous) of Monmouth County, New Jersey [abstract]. Society of Vertebrate Paleontology annual meeting, Rapid City, S.D.

Desmond, A. J. 1975. *The hot-blooded dinosaurs: A revolution in palaeontology.* London: Blond and Briggs.

Detjen, James. 1987. Dinosaurs among us. *Philadelphia Inquirer Magazine,* May 3.

Dickens, C. 1852–1853. *Bleak House.* New York: Penguin Books.

Dietrich, R. V. 1990. *Geology and Virginia.* 2d ed. Charlottesville: Virginia Division of Mineral Resources and University Press of Virginia.

Dilcher, D. L., and P. R. Crane. 1984. In pursuit of the first flower. *Natural History,* March, 56–61.

Dodson, P. 1990. Counting dinosaurs: How many kinds were there? *National Academy of Sciences Proceedings* 87:7608–7612.

REFERENCES

Dodson, P., and L. P. Tatarinov. 1990. Dinosaur extinction. In *The Dinosauria,* ed. D. B. Weishampel, P. Dodson, and H. Osmólska, 55–62. Berkeley: University of California Press.

Doyle, J. A., and L. J. Hickey. 1976. Pollen and leaves from the mid-Cretaceous Potomac Group and their bearing on early angiosperm evolution. In *Origin and early evolution of angiosperms,* ed. C. B. Beck, 139–206. New York: Columbia University Press.

Dunbar, C. O. 1949. *Historical geology.* New York: Wiley.

Edelson, Z. 1982. Observances honor O. C. Marsh. *Discovery* 16:32–36.

Edwards, A. M. 1895. Ornithichnites and jaw bone from Newark Sandstone of New Jersey. *American Journal of Science* 50:346.

Emerson, B. K., and F. B. Loomis. 1904. On *Stegomus longipes,* a new reptile from the Triassic sandstones of the Connecticut Valley. *American Journal of Science* 17:377–380.

Emmons, E. 1856. *Geological report of the midland counties of North Carolina.* North Carolina Geological Survey, 1852–1863. New York: Putnam.

———. 1857. *American geology, containing a statement of the principles of the science, with full illustrations of the characteristic American fossils, with an atlas and a geological map of the United States.* Part 6. Albany, N.Y.: Sprague.

———. 1858. *Agriculture of the eastern counties, together with descriptions of the fossils of the marl beds.* Report. Raleigh: North Carolina Geological Survey.

Eyerman, J. 1886. Footprints on the Triassic sandstone (Jura-Trias) of New Jersey. *American Journal of Science* 31:72.

———. 1889. Fossil footprints from the Jura-Trias of New Jersey. *Academy of Natural Sciences of Philadelphia Proceedings,* 32–33.

Faille, R. T. 1973. Tectonic development of the Triassic Newark-Gettysburg Basin in Pennsylvania. *Geological Society of America Bulletin* 84:725–740.

Fisher, D. W. 1958. 300 million years ago: A fishing expedition in the rocks of New York State. Reprinted from *New York State Conservationist,* February–March.

———. 1981. The world of *Coelophysis:* A New York dinosaur of 200 million years ago. *New York State Geological Survey Circular* 49:1–22.

Fitzgerald, M. 1992. Where the dinosaurs roamed in New Jersey: A marl pit in Gloucester County turns into a treasure trove for paleontologist. *Philadelphia Inquirer,* December 27, B1, B4.

Forster, C. A., and E. E. Spamer. 1986. A paleontological pilgrimage through Philadelphia, the birthplace of American paleontology, part 1. The *Mosasaur* 3:181–193.

Foulke, W. P. 1858. [Remarks concerning fossils from Haddonfield, New Jersey]. *Academy of Natural Sciences of Philadelphia Proceedings,* 213–215.

Fraser, N. C. 1988. Latest Triassic terrestrial vertebrates and their biostratigraphy. *Modern Geology* 13:125–140.

Fraser, N. C., and H.-D. Sues, eds. 1994. *In the shadow of the dinosaurs: Early Mesozoic tetrapods.* New York: Cambridge University Press.

Frederick, D. J. 1989. Rare pre-dinosaur fossils found at site in Virginia. National Geographic Society News Release, Washington, D.C., December 15.

Friis, E. M., W. G. Chaloner, and P. R. Crane, eds. 1987. *The origins of angiosperms and their biological consequences.* Cambridge: Cambridge University Press.

Froelich, A. J., and P. E. Olsen. 1984. Newark Supergroup: A revision of the Newark Group in eastern North America. *U.S. Geological Survey Bulletin* 1537A:A55–A58.

Froelich, A. J., and G. R. Robinson Jr., eds. 1988. Studies of the Early Mesozoic basins of the eastern United States. *U.S. Geological Survey Bulletin* 1776: 1–423.

Frye, K. 1986. *Roadside geology of Virginia.* Missoula, Mont.: Mountain Press.

Gallagher, W. B. 1982. Hadrosaurian dinosaurs from Delaware. *Delaware Valley Paleontological Society Newsletter* 4, no. 10:3–4.

———. 1983. Paleoecology of the Delaware Valley region. Part 1: Cambrian to Jurassic. *Mosasaur* 1:23–42.

———. 1984. Paleoecology of the Delaware Valley region. Part 2: Cretaceous to Quaternary. *Mosasaur* 2:9–44.

———. 1986a. Biostratigraphic succession and benthic paleoecology at the Cretaceous-Tertiary transition in New Jersey. North American Paleontological Convention, University of Colorado, Boulder. *Abstracts with Programs* A17.

———. 1986b. Depositional environments, paleo-oceanography, and paleoecology of the Upper Cretaceous–Lower Tertiary sequence in the New Jersey Coastal Plain. In *Geological investigations of the Coastal Plain of southern New Jersey,* part 2B, *Paleontological investigations,* ed. R. W. Talkington, 1–17. Proceedings of Symposia, 2d Annual Meeting of the Geological Society of New Jersey. Pomona, N.J.: Stockton State College.

———. 1988. Patterns of extinction and recovery across the K/T boundary in New Jersey. Geological Society of America, annual meeting, Denver, Colo. *Abstracts with Programs* 20:A106.

———. 1989a. Biostratigraphy and paleoecology of the K/T extinction in New Jersey [abstract]. *New Jersey Academy of Science Bulletin* 34:36.

———. 1989b. Comparative taphonomy of bone concentrations in the Atlantic Coastal Plain [abstract]. *Journal of Vertebrate Paleontology* 9, no. 3, Suppl.:22A.

———. 1990a. Biostratigraphy and paleoecology of the Upper Cretaceous/Lower Tertiary sequence in the New Jersey Coastal Plain. Ph.D. diss., Department of Geology, University of Pennsylvania, Philadelphia.

———. 1990b. *Dinosaurs, creatures of time.* Bulletin 14. Trenton: New Jersey State Museum.

———. 1991. Selective extinction and survival across the Cretaceous/Tertiary boundary in the northern Atlantic Coastal Plain. *Geology* 19:967–970.

———. 1992. Geochemical investigations of the Cretaceous/Tertiary boundary in the Inversand pit, Gloucester County, New Jersey. *New Jersey Academy of Science Bulletin* 37, no. 1:19–24.

———. 1993. The Cretaceous/Tertiary mass extinction event in the northern Atlantic Coastal Plain. *Mosasaur* 5:75–154.

Gallagher, W. B., and D. C. Parris. 1985. Biostratigraphic succession across the Cretaceous-Tertiary boundary at the Inversand Company pit, Sewell, New Jersey. In *Geological investigations of the coastal plain of southern new Jersey,* C1–C16. Proceedings, annual meeting of the Geological Society of New Jersey. Pomona, N.J.: Stockton State College.

Gallagher, W. B., D. C. Parris, and E. E. Spamer. 1986. Paleontology, biostratigraphy, and depositional environments of the Cretaceous-Tertiary transition in the New Jersey Coastal Plain. *Mosasaur* 3:1–35.

Gallup, M. E. 1988. A new look at the old dinosaurs of Maryland. *Maryland Naturalist* (Natural History Society of Maryland) 32, nos. 3–4:41–51.

REFERENCES

Galton, P. M. 1971. The prosauropod dinosaur *Ammosaurus,* the crocodile *Protosuchus,* and their bearing on the age of the Navajo Sandstone of northeastern Arizona. *Journal of Paleontology* 45:781–795.

———. 1976. Prosauropod dinosaurs (Reptilia: Saurischia) of North America. *Postilla,* no. 169:1–98.

———. 1983. The oldest ornithischian dinosaurs in North America from the Late Triassic of Nova Scotia, North Carolina, and Pennsylvania. Geological Society of America, annual meeting, Northeastern Section. *Abstracts with Programs* 15:122.

———. 1990. Basal Sauropodomorpha: Prosauropods. In *The Dinosauria,* ed. D. B. Weishampel, P. Dodson, and H. Osmólska, 320–344. Berkeley: University of California Press.

Galton, P. M., and J. A. Jensen. 1979. Remains of ornithopod dinosaurs from the Lower Cretaceous of North America. *Brigham Young University Geology Studies* 25:1–10.

Gauthier, J. A. 1986. Saurischian monophyly and the origin of birds. *Memoirs of the California Academy of Sciences,* no. 8:1–55.

Geikie, A. 1962. *The founders of geology.* 2d ed. New York: Dover.

Gemperlein, J. 1980. Oh, pity this lame beast of prehistory. *Philadelphia Inquirer,* October 12, 1B, 10B.

Gill, H. E., L. A. Sirkin, and J. A. Doyle. 1969. Cretaceous deltas in the New Jersey Coastal Plain. Geological Society of America, annual meeting, Atlantic City, N.J. *Abstracts with Programs* 7:79.

Gillette, D. D. 1978. Catalogue of type specimens of fossil vertebrates, Academy of Natural Sciences of Philadelphia. Part 4: Reptilia, Amphibia, and tracks. *Academy of Natural Sciences of Philadelphia Proceedings* 129:101–111.

Gillette, D. D., and M. G. Lockley, eds. 1989. *Dinosaur tracks and traces.* New York: Cambridge University Press.

Gilmore, C. W. 1919. An ornithomimid dinosaur in the Potomac of Maryland. *Science,* n.s., 50:394–395.

———. 1920. Osteology of the carnivorous Dinosauria in the United States National Museum, with special reference to the genera *Antrodemus* (*Allosaurus*) and *Ceratosaurus.* *U.S. National Museum Bulletin* 110:1–154.

———. 1921. The fauna of the Arundel Formation of Maryland. *U.S. National Museum Proceedings,* 59, no. 2389:581–594.

———. 1924. Collecting fossil footprints in Virginia. *Smithsonian Miscellaneous Collections* 76, no. 10:16–18.

———. 1928. A new fossil reptile from the Triassic of New Jersey. *U.S. National Museum Proceedings* 73, no. 7:1–8.

Glaeser, P. 1963. Lithostratigraphic nomenclature of the Newark-Gettysburg Basin. *Pennsylvania Academy of Sciences Proceedings* 37:179–188.

Glaser, J. D. 1966. *Provenance, dispersal, and depositional environments of Triassic sediments in the Newark-Gettysburg basin.* Bulletin G43. Harrisburg: Pennsylvania Geological Survey.

———. 1969. *Petrology and origin of Potomac and Magothy (Cretaceous) sediments, middle Atlantic Coastal Plain.* Report of Investigations 11. Baltimore: Maryland Geological Survey.

———. 1979. *Collecting fossils in Maryland.* Educational Series 4. Baltimore: Maryland Geological Survey. Revised and reprinted 1986.

Gore, P. J. W. 1988. Paleoecology and sedimentology of a Late Triassic lake, Culpeper Basin, Virginia. *Palaeogeography, Palaeoclimatology, Palaeoecology* 62:593–608.

Gore, P. J. W., and R. C. Lindholm. 1983. Paleoecology of Triassic and Jurassic lacustrine deposits in the Culpeper Basin of northern Virginia. Geological Society of America annual meeting, Northeastern Section. *Abstracts with Programs* 15:122.

Gould, S. J. 1980. *The panda's thumb.* New York: W. W. Norton.

Grandstaff, B., D. C. Parris, and R. K. Denton Jr. 1987. The Ellisdale local fauna (Campanian, New Jersey). *Journal of Vertebrate paleontology* 7, no. 3, suppl.:17A.

Grantham, R. G. 1989. Dinosaur tracks and megaflutes in the Jurassic of Nova Scotia. In *Dinosaur tracks and traces,* ed. D. D. Gillette and M. G. Lockley, 281–284. New York: Cambridge University Press.

Gregory, J. T. 1962. The relationships of the American phytosaur *Rutiodon. American Museum Novitates,* no. 2095:1–22.

Groot, J. J., D. M. Organist, and H. G. Richards. 1954. Marine Upper Cretaceous formations of the Chesapeake and Delaware Canal. *Delaware Geological Survey Bulletin* 3:1–64.

Guinness, E. A., Jr., and R. A. Naylor Jr. 1975. Preliminary field report on the dinosaur tract at educational park. *New Jersey Academy of Science* 20:41.

Guralnick, S. 1972. Geology and religion before Darwin: The case of Edward Hitchcock, theologian and geologist. *Isis* 63:529–543.

Haff, J. C. 1952. Memorial to Mignon Talbot. *Geological Society of America, Annual Report for 1951,* 157–158.

Hall, J. 1821. Fossil bones found in East Windsor, Connecticut. *American Journal of Science* 3:247.

Hallam, A. 1975a. *Jurassic environments.* Cambridge: Cambridge University Press.

Hammer, W. R., and W. J. Hickerson. 1994. A crested theropod dinosaur from Antarctica. *Science* 264:828–830.

Hartnagel, C. A., and S. C. Bishop. 1922. The mastodons, mammoths, and other Pleistocene mammals of New York State; being a descriptive record of all known occurrences. *New York Geological Survey Bulletins* 241, 242:1–110.

Hartstein, E. F. 1985. A new Chesapeake and Delaware Canal dredge site. *Delaware Valley Paleontological Society Newsletter* 7:3.

Hartstein, E. F., and L. E. Decina. 1986. A new Severn Formation (early middle Maastrichtian, Late Cretaceous) locality in Prince Georges County, Maryland. *Mosasaur* 3:87–95.

Hays, I. 1830. Description of a fragment of the head of a new fossil animal, discovered in a marl pit near Moorestown, New Jersey. *American Philosophical Society Transactions* 3:471–477.

Hecht, M. K., J. H. Ostrom, G. Viohl, and P. Wellnhofer, eds. 1985. *The beginnings of birds.* International *Archaeopteryx* Conference proceedings. Eichstätt, West Germany: Freund des Jura-Museums.

Heilmann, G. 1927. *The origin of birds.* New York: Appleton.

Helfrich, R. B., E. M. Dashiell, and E. D. Martin. 1936. Resolution in memory of Arthur Barneveld Bibbins. Passed at a special meeting of the Board of Directors of the Star Spangled Banner Flag House Association, Baltimore, Md., October 26.

Heron, S. D., Jr., and W. H. Wheeler. 1964. *The Cretaceous formations along the Cape Fear Riv-*

er, North Carolina. Field Conference Guidebook 5. Atlantic Coastal Plain Geological Association.

Hess, H. H. 1962. History of the ocean basins. In *Petrologic studies: A volume to honor A. F.
Buddington,* ed. A. E. J. Engel, H. L. James, and B. F. Leonard, 599–620. Boulder, Colo.:
Geological Society of America.

Hickey, L. J., and J. A. Doyle. 1977. Early Cretaceous fossil evidence for angiosperm evolution. *Botanical Review* 43:3–104.

Hickok, W. O., and B. Willard. 1933. Dinosaur foot tracks near Yocumtown, York County, Pennsylvania. *Pennsylvania Academy of Sciences Proceedings,* 7:55–58.

Hildebrand, A. R., G. T. Penfield, D. A. Kring, M. Pilkington, A. Carmargo Z., S. B. Jacob-
son, and W. V. Boynton. 1991. Chicxulub crater: A possible Cretaceous/Tertiary boundary impact crater on the Yucatán Peninsula, Mexico. *Geology* 19:867–871.

Hitchcock, C. H. 1855. Impressions (chiefly tracks) on alluvial clay in Hadley, Massachusetts. *American Journal of Science* 19:391–396.

———. 1871. Account and complete list of the ichnozoa of the Connecticut Valley. In
Walling and Gray's official topographic atlas of Massachusetts, 20–21. Boston: Walling and
Gray.

———. 1889. Recent progress in ichnology. *Boston Society of Natural History Proceedings*
24:117–127.

Hitchcock, E. B. 1836. Ornithichnology: Description of the footmarks of birds (Ornithichnites) on new red sandstone in Massachusetts. *American Journal of Science*
29:307–340.

———. 1837a. Ornithichnites in Connecticut. *American Journal of Science* 31:174–175.

———. 1837b. Fossil footsteps in sandstone and graywacke. *American Journal of Science*
32:174–176.

———. 1845. An attempt to name, classify, and describe the animals that made the fossil
footmarks of New England. *Association of American Geologists and Naturalists Proceedings,*
23–25.

———. 1851. *Religion of geology and its connected sciences.* Boston: Crosby, Nichols, Lee.

———. 1858. *Ichnology of New England: A report on the sandstone of the Connecticut Valley, especially its fossil footmarks.* Boston: William White.

———. 1865. *Supplement to the "Ichnology of New England."* Boston: Wright and Potter.

Holtz, T. R., Jr. 1994. The phylogenetic position of the Tyrannosauridae: Implications for
theropod systematics. *Journal of Paleontology* 68:1100–1117.

Hopkins, G. M. 1878. *Atlas of fifteen miles around Washington, including the County of Prince
George, Maryland.* Reprinted, ed. F. F. White Jr., Riverdale, Md.: Prince Georges County Historical Society.

Hopson, J. A. 1975. The evolution of cranial display structures in hadrosaurian dinosaurs.
Paleobiology 1:21–43.

———. 1977. Relative brain size and behavior in archosaurian reptiles. *Annual Review of
Ecology and Systematics* 8:429–448.

Horner, J. R. 1979. Upper Cretaceous dinosaurs from the Bearpaw Shale (marine) of
south-central Montana, with a checklist of Upper Cretaceous dinosaur remains from
marine sediments in North America. *Journal of Paleontology* 53:566–578.

———. 1984. The nesting behavior of dinosaurs. *Scientific American* 250:130–137.

Horner, J. R., and J. Gorman. 1988. *Digging dinosaurs.* New York: Workman.

Horner, J. R., and R. Makela. 1979. Nest of juveniles provides evidence of family structure among dinosaurs. *Nature* 282:296–298.

Horton, J. W., Jr. and V. A. Zullo, eds. 1991. *The geology of the Carolinas.* Knoxville: University of Tennessee Press.

Hoskins, D. M., J. D. Inners, and J. A. Harper. 1983. *Fossil collecting in Pennsylvania.* 3d ed. General Geology Report G40, 4th ser. Harrisburg: Pennsylvania Geological Survey.

Howard, R. W. 1975. *The dawnseekers: The first history of American paleontology.* New York: Harcourt Brace Jovanovich.

Huber, P., S. G. Lucas, and A. P. Hunt. 1993a. Revised age and correlations of the Upper Triassic Chatham Group (Deep River Basin, Newark Supergroup), North Carolina. *Southeastern Geology* 33, no. 4:171–193.

———. 1993b. Vertebrate biochronology of the Newark Supergroup Triassic, eastern North America. In *The nonmarine Triassic,* ed. S. G. Lucas and M. Morales, 179–186. Bulletin 3. Albuquerque: New Mexico Museum of Natural History and Science.

Hubert, J. F., A. A. Reed, W. L. Dowdall, and J. M. Gilchrist. 1978. *Guide to the Mesozoic redbeds of central Connecticut.* Guidebook 4. Hartford: Connecticut Geological and Natural History Survey.

Huene, F. R. von. 1913. A new phytosaur from the Palisades near New York. *American Museum of Natural History Bulletin* 32:275–283.

———. 1921. Reptilian and stegocephalian remains from the Triassic of Pennsylvania in the Cope Collection. *American Museum of Natural History Bulletin* 44:561–574.

Hunt, A. P., and S. G. Lucas 1994. Ornithischian dinosaurs from the Upper Triassic of the United States. In *In the shadow of the dinosaurs,* ed. N. C. Fraser and H.-D. Sues, 227–241. New York: Cambridge University Press.

Hunter, W. 1769. Observations on the bones commonly supposed to be elephant's bones, which have been found near the River Ohio, in America. *Royal Society of London Philosophical Transactions 1768* 58:34–45.

Huxley, T. H. 1868. On the animals which are most nearly intermediate between birds and reptiles. *Geology Magazine* 5:357–365.

———. 1870. Further evidence of the affinity between the dinosaurian reptiles and birds. *Geological Society of London Quarterly Journal* 26:12–31.

Isachsen, Y. W. 1980. *Continental collisions and ancient volcanoes: The geology of southeastern New York.* Educational Leaflet 24. Albany: New York State Geological Survey.

Isachsen, Y. W., E. Landing, J. M. Lauber, L. V. Rickard, and W. B. Rogers. 1991. *Geology of New York: A simplified account.* Educational Leaflet 28. Albany: New York State Geological Survey.

Jackson, K. L. 1989a. Nobody had time or money to save dinosaur tracks. *Allentown Morning Call,* July 5, B1.

———. 1989b. Commissioner wants to preserve dinosaur tracks. *Allentown Morning Call,* July 25, B1.

———. 1991. Historical panel backs saving dinosaur tracks. *Allentown Morning Call,* March 15, B7.

Jefferson, T. 1785. *Notes on the state of Virginia.* London: John Stockdale. Reprinted Chapel Hill: University of North Carolina Press, 1955.

―――. 1799. A memoir on the discovery of certain bones of a quadruped of the clawed kind in the western parts of Virginia. *American Philosophical Society Transactions,* o.s., 4:246–260.

Jepsen, G. L. 1948. *A Triassic armored reptile from New Jersey* (Stegomus arcuatus jerseyensis). Miscellaneous Geologic Paper 20. Trenton: New Jersey Department of Conservation.

Jerison, H. J. 1969. Brain evolution and dinosaur brains. *American Naturalist* 103:575–588.

Johnson, M. E., and D. B. McLaughlin. 1957. Triassic formations in the Delaware Valley. *Geological Society of America Guidebook for Field Trips,* no. 2:31–59.

Johnson, M. E., and H. G. Richards. 1952. Stratigraphy of the Coastal Plain of New Jersey. *American Association of Petroleum Geologists Bulletin* 36:2150–2160.

Johnston, C. 1859. Note upon odontography. *American Journal of Dental Science* 9:337–343.

Jordan, R. R. 1962. Stratigraphy of the sedimentary rocks of Delaware. *Delaware Geological Survey Bulletin* 9:1–51.

―――. 1963. Configuration of the Cretaceous-Tertiary boundary in the Delmarva peninsula and vicinity. *Southeastern Geology* 4:187–198.

―――. 1976. The Cretaceous-Tertiary boundary in Delaware. In *Guidebook to the stratigraphy of the Atlantic Coastal Plain in Delaware,* ed. A. M. Thompson, 74–80. Third annual field trip. New York: Petroleum Exploration Society of New York.

―――. 1983. *Stratigraphic nomenclature of nonmarine Cretaceous rocks of the inner margin of the coastal plain in Delaware and adjacent states.* Report of Investigations 37. Newark: Delaware Geological Survey.

Kesler, N. 1982. Corps bulldozes top geological site at the C&D Canal. *Wilmington Morning News,* February 23, A-1.

Kindle, E. M. 1931. The story of the discovery of Big Bone Lick. *Kentucky Geological Survey,* ser. 6, 16:195–212.

―――. 1935. American Indian discoveries of vertebrate fossils. *Journal of Paleontology* 9:449–452.

Kingham, R. F. 1962. Studies of the sauropod dinosaur *Astrodon* Leidy. *Washington Junior Academy of Science Proceedings* (Washington, D.C.) 1:38–44.

Klein, G. deV. 1962. Triassic sedimentation, Maritime Provinces, Canada. *Geological Society of America Bulletin* 73:1127–1146.

―――. 1968. Sedimentology of Triassic rocks in the lower Connecticut Valley. In *Guidebook for fieldtrips in Connecticut,* ed. P. M. Orville, trip C-1, sec. B-4:1–19. New England Intercollegiate Geology Conference, annual meeting, Yale University. Guidebook 2. Hartford: Connecticut Geological and Natural History Survey.

―――. 1969. Deposition of Triassic sedimentary rocks in separate basins, eastern North America. *Geological Society of America Bulletin* 80:1825–1832.

Koch, R. C., and R. K. Olsson. 1974. Microfossil biostratigraphy of the uppermost Cretaceous beds of New Jersey. Geological Society of America, annual meeting, Northeastern Section. *Abstracts with Programs* 6:45–46.

Koehler, R. 1989. Fossil friends bag dinosaur tracks. *Reading Eagle/Times,* December 3, A-1, A-2.

―――. 1990. Big Boom evidence found here. *Reading Eagle/Times,* February 25, A-1, A-6.

Kranz, P. M. 1989. *Dinosaurs in Maryland.* Educational Series, no. 6. Baltimore: Maryland Geological Survey.

Krynine, P. D. 1950. *Petrology, stratigraphy, and origin of the Triassic sedimentary rocks of Connecticut.* Bulletin 73. Hartford: Connecticut State Geological and Natural History Survey.

Kummel, H. B., and G. H. Knapp. 1904. The stratigraphy of the New Jersey clays. *New Jersey Geological Survey Final Report* 6:117–209.

Langston, W., Jr. 1960. The vertebrate fauna of the Selma Formation of Alabama. Part 6: The dinosaurs. *Fieldiana Geology Memoirs* 3:313–361.

Lanham, U. 1973. *The bone hunters.* New York: Columbia University Press.

Lauginiger, E. M. 1978. Spoil bank collecting at the Chesapeake and Delaware Canal. *Earth Science* 31:119–120.

———. 1982. Fossils and formations of the Chesapeake and Delaware Canal, Delaware. *Fossils Quarterly* 1:39–43.

———. 1984. An upper Campanian vertebrate fauna from the Chesapeake and Delaware Canal. *Mosasaur* 2:141–149.

———. 1986. An upper Cretaceous vertebrate assemblage from Big Brook, New Jersey. *Mosasaur* 3:53–61.

———. 1988. *Cretaceous fossils from the Chesapeake and Delaware Canal, a guide for students and collectors.* Special Publication 18. Newark: Delaware Geological Survey.

———. 1989. Delaware Valley Paleontological Society: "Ad amorem rerum fossam." *Mosasaur* 4:165–178.

Lauginiger, E. M., and E. F. Hartstein. 1981. Delaware fossils. *Delaware Mineralogical Society* 11:1–38.

Lea, I. 1857. Some observations of the geology of the new red sandstone formation near Gwynedd. *Academy of Natural Sciences of Philadelphia Proceedings* 9:173.

Lee, K. Y. 1980. Triassic-Jurassic geology of the northern part of the Culpeper Basin, Virginia and Maryland. Open File Report 79-1561. Washington, D.C.: U.S. Geological Survey.

Lee, K. Y., and A. J. Froelich. 1989. *Triassic-Jurassic stratigraphy of the Culpeper and Barboursville Basins, Virginia and Maryland.* Professional paper 1472. Washington, D.C.: U.S. Geological Survey.

Leidy, J. 1851. Descriptions of fossils from the greensand of New Jersey. *Academy of Natural Sciences of Philadelphia Proceedings,* 329–331.

———. 1856a. Notices on the remains of extinct vertebrate animals of New Jersey collected by Cook of the State Geological Survey under the direction of Dr. W. Kitchell. *Academy of Natural Sciences of Philadelphia Proceedings,* 221.

———. 1856b. Notices of remains of extinct vertebrate animals discovered by Prof. E. Emmons. *Academy of Natural Sciences of Philadelphia Proceedings,* 255–257.

———. 1858. [Remarks concerning *Hadrosaurus foulkii*]. *Academy of Natural Sciences of Philadelphia Proceedings,* 215–218.

———. 1859a. Account of the remains of a fossil extinct reptile recently discovered at Haddonfield, New Jersey. *Academy of Natural Sciences of Philadelphia Proceedings,* 1–16.

———. 1859b. Some remarks on some reptilian remains found near Phoenixville, Pennsylvania. *Academy of Natural Sciences of Philadelphia Proceedings,* 110.

———. 1859c. *Hadrosaurus foulkii,* a new saurian from the Cretaceous of New Jersey, related to the *Iguanodon. American Journal of Science* 27:266–270.

———. 1865. Memoir on the extinct reptiles of the Cretaceous formations of the United
States. *Smithsonian Contributions to Knowledge* 14:1–135.

Lessem, D. 1992. *Dinosaurs rediscovered: New findings which are revolutionizing dinosaur science.* New York: Touchstone Books.

Lewis, J. V., and H. B. Kummel. 1940. *The geology of New Jersey.* Geological Series, Bulletin
50. Trenton: New Jersey Department of Conservation and Development.

Lewis, M., and W. Clark. 1814. *History of the expedition under the command of Captains Lewis
and Clark, to the sources of the Missouri, thence across the Rocky Mountains and down the River
Columbia to the Pacific Ocean. Performed during the Years 1804–6. By Order of the government
of the United States.* 2 vols. Philadelphia: Paul Allen.

Lindholm, R. C. 1979. Geologic history and stratigraphy of the Triassic-Jurassic Culpeper
Basin, Virginia. *Geological Society of America Bulletin* 90, no. 11: part 1, 995–997, part 2,
1702–1736.

Little, R. D. 1984. *Dinosaurs, dunes, and drifting continents: The geohistory of the Connecticut Valley.* Greenfield, Mass.: Valley Geology Publications.

Litwin, R. J., and S. R. Ash. 1993. Revision of the biostratigraphy of the Chatham Group
(Upper Triassic), Deep River Basin, North Carolina. *Review of Paleobotany and Palynolol-
ogy* 77:75–95.

Litwin, R. J., A. Traverse, and S. R. Ash. 1991. Preliminary palynological zonation of the
Chinle Formation, southwestern USA, and its correlation to the Newark Supergroup
(eastern USA). *Review of Paleobotany and Palynology* 68:269–287.

Litwin, R. J., and R. E. Weems. 1992. Re-evaluation of the age of Triassic strata (Doswell
Formation) of the Taylorsville Basin, VA. *Virginia Journal of Science* 43:265.

Lockley, M. G. 1991. *Tracking dinosaurs: A new look at an ancient world.* New York: Cambridge
University Press.

Lucas, F. A. 1926. Thomas Jefferson: Paleontologist. *Natural History* 26, no. 3:328–330.

Lucas, S. G., and P. Huber. 1993. Revised internal correlation of the Newark Supergroup
Triassic, eastern United States and Canada. In *The nonmarine Triassic,* ed. S. G. Lucas and
M. Morales. 311–319. Bulletin no. 3. Albuquerque: New Mexico Museum of Natural
History and Science.

Lucas, S. G., and M. Morales, eds. 1993. *The nonmarine Triassic.* Bulletin no. 3. Albuquerque: New Mexico Museum of Natural History and Science.

Lull, R. S. 1904. Fossil footprints of the Jura-Trias in North America. *Boston Society of Natural History Memoirs* 5:461–557.

———. 1911a. The reptilian fauna of the Arundel Formation. In *Maryland Geological Survey Lower Cretaceous* 173–180. Baltimore: Johns Hopkins Press.

———. 1911b. Systematic paleontology, Lower Cretaceous: Vertebrata. In *Maryland Geological Survey Lower Cretaceous,* 183–210. Baltimore: Johns Hopkins Press.

———. 1912. The life of the Connecticut Trias. *American Journal of Science* 33:397–422.

———. 1915. *Triassic life of the Connecticut Valley.* Bulletin 24. Hartford: Connecticut Geological and Natural History Survey.

———. 1953. *Triassic life of the Connecticut River Valley.* Rev. ed. Bulletin 81. Hartford: Connecticut Geological and Natural History Survey.

Lull, R. S., and N. E. Wright. 1942. Hadrosaurian dinosaurs of North America. Special Paper 40. Boulder, Colo.: Geological Society of America.

Luttrell, G. W. 1989. *Stratigraphic nomenclature of the Newark Supergroup of eastern North America.* Bulletin 1572. Washington, D.C.: U.S. Geological Survey.

Lyell, C. 1843. On the fossil footprints of birds and impressions of raindrops in the valley of the Connecticut. *American Journal of Science* 45:394–397.

———. 1845. The Cretaceous strata of New Jersey. *Geological Society of London Proceedings* 1:55–60.

Maceo, P. J., and D. H. Riskind. 1989. Field and laboratory moldmaking and casting of dinosaur tracks. In *Dinosaur tracks and traces,* ed. D. D. Gillette and M. G. Lockley, 419–420. New York: Cambridge University Press.

Mansfield, G. R. 1922. Potash in greensands of New Jersey. *U.S. Geological Survey Bulletin* 727:5–12.

Mantell, G. A. 1825. Notice on the *Iguanodon,* a newly discovered fossil reptile from the sandstone of Tilgate Forest in Sussex. *Royal Society of London Philosophical Transactions* 115:179–186.

———. 1845. Description of footmarks and other imprints on a slab of new red sandstone, from Turner's Falls, Massachusetts, U.S., collected by Dr. James Deane of Greenfield, U.S *Geological Society of London Quarterly Journal* 2:38.

Marsh, O. C. 1870a. A new dinosaurian from the Cretaceous green sand near Barnsboro, New Jersey. *Academy of Natural Sciences of Philadelphia Proceedings* 22:2.

———. 1870b. [Remarks on *Hadrosaurus minor, Mosasaurus crassidens, Leiodon laticaudus, Baptosaurus,* and *Rhinoceros matutinus*]. Academy of Natural Sciences of Philadelphia Proceedings 22:2–3.

———. 1872. Notice on a new species of *Hadrosaurus. American Journal of Science* 3:301.

———. 1888. Notice of a new genus of Sauropoda and other new dinosaurs from the Potomac Formation. *American Journal of Science* 35:89–94.

———. 1893. Restoration of *Anchisaurus. American Journal of Science* 45:169–170.

———. 1896. A New Belodont reptile (*Stegomus*) from the Connecticut River sandstone. *American Journal of Science* 2:59–62.

———. 1898. Jurassic Formation on the Atlantic coast. *American Journal of Science* 6, suppl.:105–115.

———. 1899. Footprints of Jurassic dinosaurs. *American Journal of Science* 7:227–232.

Mathews, W. H. 1989. Pioneers of paleontology. *Earth Science* 42 (winter): 17–19.

Mattis, A. F. 1975. Early Mesozoic rifting and sedimentation, Morocco and eastern North America. *New Jersey Academy of Science Abstract* 20, no. 1:41.

McDonald, N. G. 1982. Paleontology of the Mesozoic rocks of the Connecticut Valley. In *Guidebook for fieldtrips in Connecticut and south central Massachusetts,* ed. R. Joesten and S. Quarrier, 143–172. New England Intercollegiate Geological Conference, annual meeting. Hartford: Connecticut Geological and Natural History Survey.

McGee, W. J. 1886. Geological formations underlying Washington and vicinity. *American Journal of Science* 31:473–474.

McGowan, C. 1991. *Dinosaurs, spitfires, and sea dragons.* Cambridge: Harvard University Press.

McIntosh, J. S. 1990. Sauropoda. In *The Dinosauria,* ed. D. B. Weishampel, P. Dodson, and H. Osmólska, 345–401. Berkeley: University of California Press.

REFERENCES

McKenna, M. C. 1962. Collecting small fossils by washing and screening. *Curator* 5:221–235.

McLaughlin, D. B. 1959. Mesozoic rocks. In *Geology and mineral resources of Bucks County, Pennsylvania*, ed. B. Willard, 55–162. Bulletin C9. Harrisburg: Pennsylvania State Geological Survey.

McLennan, J. D. 1973. *Dinosaurs in Maryland.* Baltimore: Maryland Geological Survey.

Merrill, G. P. 1924. *The first one hundred years of American Geology.* New Haven: Yale University Press.

Meyer, E. L. 1992. DinoMite dinosaur park proposed for Prince George's. *Washington Post,* July 18, B3.

Meyertons, C. T. 1963. *Triassic formations of the Danville Basin.* Report of Investigations 6. Charlottesville: Virginia Division of Mineral Resources.

Miller, C. T. 1979. Delaware's Cretaceous fossils. *Earth Science* 24:247–251.

Miller, H. W., Jr. 1955. A checklist of the Cretaceous and Tertiary vertebrates of New Jersey. *Journal of Paleontology* 29:903–914.

———. 1962. The Cretaceous reptiles of New Jersey. In *The Cretaceous fossils of New Jersey, part 2,* ed. H. G. Richards et al., 193–196. Trenton, N.J.: Bureau of Geology and Topography, Department of Conservation and Economic Development.

———. 1967. Cretaceous vertebrates from Phoebus Landing, North Carolina. *Academy of Natural Sciences of Philadelphia Proceedings* 119:219–235.

———. 1968. Additions to the Upper Cretaceous vertebrates from Phoebus Landing, North Carolina. *Elisha Mitchell Scientific Society Journal* 84:467–471.

Minard, J. P., J. P. Owens, N. F. Sohl, H. E. Gill, and J. F. Mello. 1969. *Cretaceous-Tertiary boundary in New Jersey, Delaware, and eastern Maryland.* Bulletin 1274-H. Washington, D.C.: U.S. Geological Survey.

Minard, J. P., N. F. Sohl, and J. P. Owens. 1978. *Re-introduction of the Severn Formation (Upper Cretaceous) to replace the Monmouth Formation in Maryland.* Bulletin 1435-A. Washington, D.C.: U.S. Geological Survey.

Mitchell, J. A. 1895. The discovery of fossil tracks in the Newark System (Jura-Trias) of Frederick County, Maryland. *Johns Hopkins University Circular* 15, no. 121: 15–16.

Moody, R. 1980. *The prehistoric world: The 3,400 million years before modern man.* Secaucus, N.J.: Chartwell Books.

Moore, J. E., and J. A. Jackson, eds. 1989. *Geology, hydrology, and history of the Washington, D.C., area.* Alexandria, Va.: American Geological Institute.

Morton, S. G. 1834. *Synopsis of the organic remains of the Cretaceous groups of the United States, illustrated by 19 plates, to which is added an appendix containing a tabular view of the Tertiary fossils hitherto discovered in North America.* Philadelphia: Key and Biddle.

Nelson, R. H. 1965. New locality for dinosaur tracks in Connecticut. *Rocks and Minerals* 40:5–7.

Newberry, J. S. 1888. *Fossil fishes and fossil plants of the Triassic rocks of New Jersey and the Connecticut Valley.* Monograph 14. Washington, D.C.: U.S. Geological Survey.

Newman, W. L. 1988. *Geologic time.* Inf-70-1. Washington, D.C.: U.S. Geological Survey.

Norman, D. B. 1985. *The illustrated encyclopedia of dinosaurs.* New York: Crescent Books.

———. 1991. *Dinosaur!* New York: Prentice-Hall.

Norman, D. B., and D. B. Weishampel. 1985. Ornithopod feeding mechanisms: Their bearing on the evolution of herbivory. *American Naturalist* 126:151–164.

Olsen, P. E. 1975. The microstratigraphy of the Roseland Quarry (Early Jurassic, Newark Supergroup, New Jersey). Open Report to the Essex County Park Commission.

———. 1978. On the use of the term Newark for Triassic and Early Jurassic rocks of eastern North America. *Newsletter of Stratigraphy* 7:90–95.

———. 1980a. Fossil great lakes of the Newark Supergroup in New Jersey. In *Field studies in New Jersey geology and guide to field trips,* ed. W. Manspeizer, 352–398. New York State Geological Association annual meeting. Newark, N.J.: Newark College of Arts and Sciences, Rutgers University.

———. 1980b. Triassic and Jurassic formations of the Newark Basin. In *Field studies in New Jersey geology and guide to field trips,* ed. W. Manspeizer, 2–39. New York State Geological Association annual meeting. Newark, N.J.: Newark College of Arts and Sciences, Rutgers University.

———. 1980c. A comparison of the vertebrate assemblages from the Newark and Hartford (Early Mesozoic, Newark Supergroup) of eastern North America. In *Aspects of vertebrate history,* ed. L. L. Jacobs, 35–54. Flagstaff, Ariz.: Museum of Northern Arizona Press.

———. 1986a. A 40-million-year lake record of early Mesozoic orbital climatic forcing. *Science* 234:842–848.

———. 1986b. Discovery of earliest Jurassic reptile assemblages from Nova Scotia imply catastrophic end to the Triassic. *Lamont Log* (Lamont-Doherty Geological Observatory, Columbia University), spring, 1–3.

———. 1988. Continuity of strata in the Newark and Hartford Basins. In *Studies of the Early Mesozoic basins of the eastern United States,* ed. A. J. Froelich and G. R. Robinson, Jr., 6–18. Bulletin 1776. Washington, D.C.: U.S. Geological Survey.

Olsen, P. E., and D. Baird. 1982. Early Jurassic vertebrate assemblages from the McCoy Brook Formation of the Fundy Group (Newark Supergroup, Nova Scotia, Canada). Geological Society of America, annual meeting. *Abstracts with Programs* 14(1/2):70.

———. 1986. the ichnogenus *Atreipus* and its significance for Triassic biostratigraphy. In *The beginning of the age of dinosaurs,* ed. K. Padian, 61–87. New York: Cambridge University Press.

Olsen, P. E., and J. J. Flynn. 1989. Field guide to the vertebrate paleontology of Late Triassic age rocks in the southwestern Newark Basin (Newark Supergroup, New Jersey and Pennsylvania). *Mosasaur* 4:1–43.

Olsen, P. E., and P. M. Galton. 1977. Triassic-Jurassic tetrapod extinctions: Are they real? *Science* 197:983–986.

Olsen, P. E., and D. V. Kent. 1990. Continental coring of the Newark Rift. *EOS* 71:385.

Olsen, P. E., A. R. McCune, and K. S. Thomson. 1982. Correlation of the Early Mesozoic Newark Supergroup by vertebrates, principally fishes. *American Journal of Science* 282:1–44.

Olsen, P. E., C. L. Remington, B. Cornet, and K. S. Thomson. 1978. Cyclic change in Late Triassic lacustrine communities. *Science* 201:729–733.

Olsen, P. E., R. W. Schlische, and P. J. W. Gore. 1989. *Field guide to the tectonics, stratigraphy, sedimentology, and paleontology of the Newark Supergroup, eastern North America.* Field Trip Guidebook T351. Washington, D.C.: International Geological Congress.

Olsen, P. E., N. H. Shubin, and M. H. Anders. 1987. New Early Jurassic tetrapod assemblages constrain Triassic-Jurassic Tetrapod extinction event. *Science* 237:1025–1029.

REFERENCES

Olsen, P. E., and H.-D. Sues. 1986. Correlation of continental Late Triassic and Early Jurassic sediments, and patterns of the Triassic-Jurassic tetrapod transition. In *The beginning of the age of dinosaurs*, ed. K. Padian, 321–351. New York: Cambridge University Press.

Olson, S. L., and D. D. Gillette. 1978. Catalogue of type specimens of fossil vertebrates, Academy of Natural Sciences, part 3, Birds. *Academy of Natural Sciences of Philadelphia Proceedings* 129:99–100.

Olson, S. L., and D. C. Parris, 1987. The Cretaceous birds of New Jersey. *Smithsonian Contributions to Paleobiology* 63:1–22.

Olsson, R. K., T. G. Gibson, H. J. Hansen, and J. P. Owens. 1988. Geology of the northern Atlantic Coastal Plain: Long Island to Virginia. In *The geology of North America: The Atlantic continental margin,* ed. R. E. Sheridan and J. A. Grow, 1–2:87–105. Boulder, Colo.: Geological Society of America.

Osborn, H. F. 1919. Biographical memoir of Joseph Leidy, 1823–1891. *National Academy of Sciences Biographical Memoirs* 8:339–396.

———. 1923. Mastodons of the Hudson Highlands. *Natural History* 23:3–24.

———. 1931. *Cope: Master naturalist.* Princeton: Princeton University Press.

———. 1935. Thomas Jefferson as a paleontologist. *Science* 82:533–538.

Ostrom, J. H. 1967a. On the discovery of *Hypsognathus* in Connecticut. *Discovery* 3:59.

———. 1967b. Peabody paleontologists assist new dinosaur track park. *Discovery* 2, no. 2:21–24.

———. 1968. The Rocky Hill dinosaurs. In *Guidebook for fieldtrips in Connecticut,* ed. P. M. Orville, trip C-3, sec. C-3, 1–12. New England Intercollegiate Geological Conference annual meeting, Yale University. Hartford: Connecticut Geological and Natural History Survey.

———. 1969a. On preparing *Hypsognathus* from Connecticut. *Discovery* 5:126.

———. 1969b. *Osteology of Deinonychus antirrhopus, an unusual theropod from the Lower Cretaceous of Montana.* Bulletin 30. New Haven: Yale Peabody Museum of Natural History.

———. 1969c. The case of the missing specimen. *Discovery* 5:50–51.

———. 1972. Were some dinosaurs gregarious? *Palaeogeography, Palaeoclimatology, Palaeoecology* 11:287–301.

———. 1976. *Archaeopteryx* and the origin of birds. *Biological Journal of the Linnean Society of London* 8:81–182.

———. 1980. The evidence for endothermy in dinosaurs. In *A cold look at the warm-blooded dinosaurs,* ed. R. D. K. Thomas and E. C. Olson, 15–54. Boulder, Colo.: Westview Press.

Owen, R. 1842. Report on British fossil reptiles, part 2. *Report of the British Association for the Advancement of Science,* 11:60–204.

———. 1849. Notes of remains of fossil reptiles discovered by Professor Henry Rodgers of Pennsylvania, U.S., in greensand formations of New Jersey. *Geological Society of London Proceedings* 5:380–383.

———. 1878. On the occurrence in North America of rare extinct vertebrates found fragmentarily in England. *Annual Magazine of Natural History* 2, no. 5:201–223.

Owens, J. P., and N. F. Sohl. 1969. Shelf and deltaic paleo-environments in the Cretaceous-Tertiary formations of the New Jersey Coastal Plain. In *Geology of selected areas in New Jersey and eastern Pennsylvania and guidebook of excursions,* ed. S. Subitzky, 253–278. New Brunswick: Rutgers University Press.

———. 1989. *Campanian and Maastrichtian depositional systems of the Black Creek Group of the Carolinas.* Durham, NC.: Guidebook. Carolina Geological Society.

Padian, K., ed. 1986a. *The beginning of the age of dinosaurs.* New York: Cambridge University Press.

———, ed. 1986b. The origin of birds and the evolution of flight. *California Academy of Science Memoirs,* no. 8:1–98.

———. 1986c. On the track of the dinosaurs. *Palaios* 1:519–520.

Padian, K., and W. A. Clemens. 1985. Terrestrial vertebrate diversity: Episodes and insights. In *Phanerozoic diversity factors,* ed. J. W. Valentine, 41–96. Princeton: Princeton University Press.

Padian, K., and C. L. May. 1993. The earliest dinosaurs. In *The nonmarine Triassic,* ed. S. G. Lucas and M. Morales, 379–381. Bulletin 3. Albuquerque: New Mexico Museum of Natural History and Science.

Pannell, N. K. 1985. Dinosaur footprints at Oak Hill, Virginia. Master's thesis, Department of Geology, George Washington University, Washington, D.C.

Parker, W. K. 1864. Remarks on the skeleton of the *Archaeopteryx;* and on the relations of the bird to the reptile. *Geological Magazine* 1:55–57.

Parris, D. C. 1986. *Biostratigraphy of the fossil crocodile* Hyposaurus owen *from New Jersey.* Investigation 4. Trenton: New Jersey State Museum.

Parris, D. C., and B. S. Grandstaff. 1989. Nonmarine microvertebrates of the Ellisdale local fauna: Campanian of New Jersey [abstract]. *Journal of Vertebrate Paleontology* 9, no. 3 suppl.: 35A.

Parris, D. C., B. S. Grandstaff, R. K. Denton Jr., W. B. Gallagher, C. DeTemple, S. S. Albright, E. E. Spamer, and D. Baird. 1987. Taphonomy of the Ellisdale dinosaur site, Cretaceous of New Jersey. Final Report on National Geographic Society Grant 3299–86.

Peale, R. 1803. *Historical disquisition on the mammoth, or great American incognitum, an extinct, immense, carnivorous animal, whose fossil remains have been found in North America.* London: C. Mercier.

Perry, K. P. 1937. Dinosaur tracks in Connecticut: Footprints in the sands of time. *Rocks and Minerals* 12:74–75.

Pickett, T. E. 1972. *Guide to common Cretaceous fossils of Delaware.* Report of Investigations 21. Newark: Delaware Geological Survey.

———. 1975. *Selected fossil collecting locations in Delaware and minerals in Delaware.* Special Publication 7. Newark: Delaware Geological Survey.

Pickett, T. E., and D. Windish. 1980. Delaware, its rocks, minerals, and fossils. Special Publication 19. Newark: Delaware Geological Survey. Revised 1992.

Powell, B. W. 1956. Dinosaur tracks of the Connecticut Valley. *Rocks and Minerals* 31:3–8.

Rader, E. K. 1964. Guide to fossil collecting in Virginia. Information Circular 7. Charlottesville: Virginia Division of Mineral Resources.

Ramsdell, R. C. 1958. Historical review of previous work on the Cretaceous of New Jersey. In *The Cretaceous fossils of New Jersey, part 1,* ed. H. G. Richards et al., 3–7. Paleontology Series, Bulletin 61. Trenton, N.J.: Bureau of Geology and Topography, Department of Conservation and Economic Development.

Raup, D. M. 1986. *The nemesis affair: A story of the death of dinosaurs and the way of science.* New York: Norton.

REFERENCES

Reed, J. C., Jr. 1989. The geology beneath Washington, D.C.: The foundations of a nation's capital. In *Geology, hydrology, and history of the Washington, D.C., area,* ed. J. E. Moore and J. A. Jackson, 27–50. Alexandria, Va.: American Geological Institute.

Richards, H. G. 1950. Geology of the Coastal Plain of North Carolina. *American Philosophical Society Transactions,* n.s., 40, pt. 1:1–83.

———. 1956. *Geology of the Delaware Valley.* Philadelphia: Mineralogical Society of Pennsylvania.

———. 1966. Philadelphia's fossils: Dinosaurs in your backyard. *Frontiers* 31:36–41.

———. 1967. Stratigraphy of the Atlantic Coastal Plain between Long Island and Georgia. *American Association of Petroleum Geologists Bulletin* 51:2400–2429.

Richards, H. G., and W. B. Gallagher. 1974. *The problem of the Cretaceous-Tertiary boundary in New Jersey.* Notulae Naturae, 449. Philadelphia: Academy of Natural Sciences of Philadelphia.

Richards, H. G., J. J. Groot, and R. M. Germeroth. 1957. Cretaceous and Tertiary geology of New Jersey, Delaware, and Maryland. In *Guidebook for field trips,* 183–230. Field Trip 6. Boulder, Colo.: Geological Society of America.

Richards, H. G., R. C. Ramsdell, B. F. Howell, J. W. Wells, and C. W. Cooke. 1958. *The Cretaceous fossils of New Jersey, part 1.* Paleontology Series, Bulletin 61. Trenton, N.J.: Bureau of Geology and Topography, Department of Conservation and Economic Development.

Richards, H. G., R. C. Ramsdell, A. K. Miller, H. F. Garner, J. B. Reeside Jr., J. A. Jeletzky, H. B. Roberts, and H. W. Miller Jr. 1962. *The Cretaceous fossils of New Jersey, part 2.* Paleontology Series, Bulletin 61. Trenton, N.J.: Bureau of Geology and Topography, Department of Conservation and Economic Development.

Ricqlès, A. J. de. 1980. Tissue structures of dinosaur bone: Functional significance and possible relation to dinosaur physiology. In *A cold look at the warm-blooded dinosaurs,* ed. R. D. K. Thomas and E. C. Olson, 103–139. Boulder, Colo.: Westview Press.

Ries, H. 1902. Report on the clays of Maryland. In *Maryland Geological Survey,* 4, part 3:203–505. Baltimore: Johns Hopkins University Press.

Ries, H., H. B. Kummel, and G. N. Knapp. 1904: *The clays and clay industry of New Jersey.* Final report 6, part 2. Trenton: Geological Survey of New Jersey.

Robb, A. J. 1989. The Upper Cretaceous (Campanian, Black Creek Formation) fossil fish fauna of Phoebus Landing, Bladen County, North Carolina. *Mosasaur* 4:75–92.

Robbins, E. I. 1991. *Age of Early Cretaceous palynomorphs in the Muirkirk clay pit fossil locality (Prince Georges County, Maryland).* Open File Report 91-613. Washington, D.C.: U.S. Geological Survey.

Roberts, D. E. 1895. Note on the Cretaceous formations of the Eastern Shore of Maryland. *Johns Hopkins University Circular,* 15, no. 121:16.

Roberts, J. K. 1928. *The geology of the Virginia Triassic.* Bulletin 29. Charlottesville: Virginia Geological Survey.

Rogers, H. D., L. Vanuxem, R. D. Taylor, E. Emmons, and T. A. Conrad. 1841. Report on the ornithichnites or foot marks of extinct birds in the new red sandstone of Massachusetts and Connecticut, observed and described by Professor Hitchcock of Amherst. *American Journal of Science* 41:165–168.

Russell, D. A. 1989. *An odyssey in time: The dinosaurs of North America.* Minocqua, Wisc.: NorthWord Press.

Russell, I. C. 1878. On the physical history of the Triassic Formation in New Jersey and the Connecticut Valley. *New York Academy of Science Annals* 1:220–254.

Russell, L. S. 1930. Upper Cretaceous dinosaur faunas of North America. *American Philosophical Society Proceedings* 69:133–159.

Russell-Robinson, S. 1989. Dinosaur tracks find home in Reston, Virginia. U.S. Geological Survey News Release, June 28.

Ryan, J. D. 1980. Triassic fossil reptile footprints near Coopersburg, Lehigh County, Pennsylvania. *Pennsylvania Geology* 11, no. 6:2–4.

Ryan, J. D., and B. Willard. 1947. Triassic footprints from Bucks County, Pennsylvania. *Pennsylvania Academy of Science Proceedings* 21:91–93.

Salter, R. 1990. Geovandalism: What happened at Coopersburg causing concern. *Allentown Morning Call,* March 13, D1.

Sanders, J. E. 1963. Late Triassic tectonic history of the northeastern United States. *American Journal of Science* 261:501–524.

Schmidt, M. F., Jr. 1993. Maryland's geology. Centreville, Md.: Tidewater.

Schoffstall, M. 1939. Finds rare dinosaur tracks in rock near Schwenksville: Earl L. Poole, assistant director of museum, proves giant *Otozoum* and tiny *Anchisauripus sillimani* once foraged for food in tidal mud near Berks. *Reading Eagle/Times,* March 9, A1, A5.

Schuchert, C., and C. M. LeVene. 1940. *O. C. Marsh: Pioneer in paleontology.* New Haven: Yale University Press.

Schulz, G., and R. C. Hope. 1973. Late Triassic microfossil flora from the Deep River Basin, North Carolina. *Palaeontographica* 141B:63–88.

Sereno, P. C. 1986. Phylogeny of the bird-hipped dinosaurs (order Ornithischia). *National Geography Research* 2:234–256.

Shaler, N. S., and J. B. Woodworth. 1899. Geology of the Richmond Basin, Virginia. *U.S. Geological Survey 19th Annual Report* 2:393–515.

Shubin, N. H., A. W. Crompton, H.-D. Sues, and P. E. Olsen. 1991. New fossil evidence on the sister-group of mammals and early Mesozoic faunal distributions. *Science* 251:1063–1065.

Silliman, B. 1837. Ornithichnites in Connecticut. *American Journal of Science* 31:165.

———. 1843. Ornithichnites of the Connecticut River sandstone and the discoveries of New Zealand, containing Dr. Deane's correspondence with Dr. Mantell. *American Journal of Science* 45:177–185.

Silvestri, S. M., and M. J. Szajna. 1993. Biostratigraphy of vertebrate footprints in the Late Triassic section of the Newark Basin, Pennsylvania: Reassessment of stratigraphic ranges. In *The nonmarine Triassic,* ed. S. G. Lucas and M. Morales, 439–445. Bulletin 3. Albuquerque: New Mexico Museum of Natural History and Science.

Simpson, G. G. 1942. The beginnings of vertebrate paleontology in North America. *American Philosophical Society Proceedings* 86:130–188.

Sinclair, W. J. 1917. A new labyrinthodont from the Triassic of Pennsylvania. *American Journal of Science* 43:319–321.

———. 1918. A large parasuchian from the Triassic of Pennsylvania. *American Journal of Science* 45:457–462.

Singewald, J. T., Jr. 1911. *Report on the iron ores of Maryland, with an account of the iron industry.* Special Publication 9, part 3. Baltimore: Maryland Geological and Economic Survey.

Smith, D., and Galton, P. 1990. Osteology of *Archaeornithomimus asiasticus* (Upper Cretaceous, Iren Dabasu Formation, People's Republic of China). *Journal of Vertebrate Paleontology* 10:255–265.

Smith, J. R. 1982. Dinosaurs in Virginia: Evidence of two new genera. *Lapidary Journal,*
September, 1110–1111.

Smith, N. 1820. Fossil bones found in red sandstone. *American Journal of Science* 2:146–147.

Sohl, N. F., and R. A. Christopher. 1983. *The Black Creek–Peedee formation contact (Upper Cretaceous) in the Cape Fear region of North Carolina.* Professional paper 1285. Washington,
D.C.: U.S. Geological Survey.

Sohl, N. F., and J. P. Owens. 1991. The Cretaceous strata of the Carolina Coastal Plain.
In *The geology of the Carolinas,* ed. J. W. Horton Jr. and V. A. Zullo, 191–220. Carolina Geological Society 50th Anniversary Volume. Knoxville: University of Tennessee Press.

Spamer, E. E. 1989. Notes on six real and supposed type fossils from the Newark Supergroup (Triassic) of Pennsylvania. *Mosasaur* 4:49–52.

Spamer, E. E., and C. A. Forster. 1989. A paleontological pilgrimage through Philadelphia,
the birthplace of American paleontology, including notes on the paleontology of
Philadelphia, part 2. *Mosasaur* 4:153–163.

Spangler, W. B., and J. Peterson. 1950. *Geology of the Atlantic Coastal Plain in New Jersey,
Delaware, Maryland, and Virginia.* Bulletin 34, no. 1. Tulsa, Okla.: American Association
of Petroleum Geologists.

Steinbock, R. T. 1989. Ichnology of the Connecticut Valley: A vignette of American science in the early nineteenth century. In *Dinosaur tracks and traces,* ed. D. D. Gillette and
M. G. Lockley, 27–32. New York: Cambridge University Press.

Stephenson, L. W. 1923. *The Cretaceous formations of North Carolina.* Raleigh: North Carolina Geological and Economic Survey.

Stuckey, J. L. 1965. *North Carolina: Its geology and mineral resources.* Raleigh: North Carolina Department of Conservation and Development.

Sues, H.-D. 1992. A remarkable new armored archosaur from the Upper Triassic of Virginia. *Journal of Vertebrate Paleontology* 12:142–149.

Sues, H.-D., and P. E. Olsen. 1990. Triassic vertebrates of Gondwanan aspect from the Richmond Basin of Virginia. *Science* 249:1020–1023.

Sues, H.-D., N. H. Shubin, and P. E. Olsen. 1994. A new sphenodontian (Lepidosauria:
Rhynchocephalia) from the McCoy Brook Formation (Lower Jurassic of Nova Scotia).
Journal of Vertebrate Paleontology 14:327–340.

Tagg, A. R., and E. Uchupi. 1966. *Distribution and geologic structure of Triassic rocks in the Bay
of Fundy and the northern part of the Gulf of Maine.* Professional paper 550B. Washington,
D.C.: U.S. Geological Survey.

Talbot, Mignon. 1911. *Podokesaurus holyokensis,* a new dinosaur from the Triassic of the
Connecticut Valley. *American Journal of Science* 31:469–479.

Thayer, P. A. 1970. Stratigraphy and geology of the Dan River Triassic Basin, North Carolina. *Southeastern Geology* 12:1–31.

Thomas, R. D. K., and E. C. Olson, eds. 1980. *A cold look at the warm-blooded dinosaurs.* Boulder, Colo.: Westview Press.

Thulborn, R. A. 1990. *Dinosaur tracks.* London: Chapman and Hall.

Thurston, H. 1986. The fossils of Fundy. *Atlantic Insight,* July, 20–23.

———. 1994. *Dawning of the dinosaurs.* Halifax, N.S.: Nimbus and the Nova Scotia Museum.

Traverse, A. 1987. Pollen and spores date origin of rift basins from Texas to Nova Scotia as early Late Triassic. *Science* 236:1469–1472.

Valentine, J. W. 1982. Darwin's impact on paleontology. *BioScience,* June, 513–518.

Van Houten, F. B. 1964. Cyclic lacustrine sedimentation, Upper Triassic Lockatong Formation, central New Jersey and adjacent Pennsylvania. In *Symposium on cyclic sedimentation,* ed. D. F. Mirriam, 11:495–531. Bulletin 169. Lawrence: Kansas Geological Survey.

———. 1965. Composition of Triassic Lockatong and associated formations of Newark Group, central New Jersey and adjacent Pennsylvania. *American Journal of Science* 263:825–863.

———. 1969. Late Triassic Newark Group, north-central New Jersey and adjacent Pennsylvania. In *Geology of selected areas in new Jersey and eastern Pennsylvania and guidebook of excursions,* ed. S. Subitzki, 314–347. Geological Society of America 1969 annual meeting. New Brunswick: Rutgers University Press.

Vokes, H. E. 1949. Maryland dinosaurs. *Maryland Naturalist* (Natural History Society of Maryland) 19, no. 3:37–46.

———. 1974. *Geography and geology of Maryland.* Rev. J. Edwards Jr. Bulletin 19. Maryland Geological Survey. Originally published 1957.

Wanner, A. 1889. The discovery of fossil tracks, algae, etc., in the Triassic of York County, Pennsylvania. *Pennsylvania Geological Survey Annual Report for 1887,* 21–35.

Ward, L. F. 1895. The Potomac Formation. *U.S. Geological Survey Annual Report,* 307–397.

Weems, R. E. 1980. An unusual newly discovered archosaur from the Upper Triassic of Virginia, USA. *American Philosophical Society Transactions* 70:1–53.

———. 1987. A Late Triassic footprint fauna from the Culpeper Basin, northern Virginia (USA). *American Philosophical Society Transactions* 77, pt. 1:1–79.

———. 1992. A re-evaluation of the taxonomy of Newark Supergroup saurischian dinosaur tracks, using extensive statistical data from a recently exposed tracksite near Culpepper, Virginia. *Virginia Division of Mineral Resources Publication* 119:113–127.

———. 1993a. *Stratigraphic distribution and bibliography of fossil fish, amphibians, and reptiles from Virginia.* Open-File Report 93-222. Washington, D.C.: U.S. Geological Survey.

———. 1993b. Upper Triassic reptile footprints and a coelacanth fish scale from the Culpeper Basin, Virginia. *Biological Society of Washington Proceedings* 106:390–401.

Weems, R. E., and C. R. Wiggs. 1991. Parasuchian occurrences in upper Triassic rocks of the Culpeper Basin of Virginia and Maryland. *Virginia Journal of Science* 42:222.

Weishampel, D. B. 1984. The evolution of jaw mechanisms in ornithopod dinosaurs. *Advances in Anatomy, Embryology and Cell Biology* 87:1–116.

———. 1990. Dinosaurian distributions. In *The Dinosauria,* ed. D. B. Weishampel, P. Dodson, and H. Osmólska, 63–139. Berkeley: University of California Press.

Weishampel, D. B., P. Dodson, and H. Osmólska, eds. 1990. *The Dinosauria.* Berkeley: University of California Press.

Weller, S. 1904. The classification of the Upper Cretaceous formations and faunas of New Jersey. In *Annual Report of the State Geologist for 1904,* 1–155. Trenton: Geological Survey of New Jersey.

————. 1907. *A Report on the Cretaceous paleontology of New Jersey.* Paleontology series, vol. 4, pt. 1. Trenton: Geological Survey of New Jersey.

Wheatley, C. M. 1861. Remarks on the Mesozoic red sandstone of the Atlantic Slope and notice of a bone bed therein, at Phoenixville, Pennsylvania. *American Journal of Science* 32:41–48.

Widmer, K. 1964. *The geology and geography of New Jersey.* Princeton, N.J.: D. Van Nostrand.

Willard, B. 1934. Additional Triassic dinosaur tracks from Pennsylvania. *Science,* n.s., 80, no. 2064:73–74.

————. 1940. Manus impressions of *Anchisauripus* from Pennsylvania. *Pennsylvania Academy of Sciences Proceedings* 14:37–39.

Wisnowsky, J. 1986. A Rosetta Stone of evolution (Nova Scotia). *Time,* February 17, 92.

Wolfe, J. A. 1976. *Stratigraphic distribution of some pollen types from the Campanian and lower Maestrichtian rocks (Upper Cretaceous) of the middle Atlantic states.* Professional paper 977. Washington, D.C.: U.S. Geological Survey.

Wolfe, P. E. 1977. *The geology and landscapes of New Jersey.* New York: Crane, Russak.

Woodworth, J. B. 1895. Three-toed dinosaur tracks in the Newark Group at Avondale, New Jersey. *American Journal of Science* 50:480–482.

Woolman, L. 1897. Bones of a dinosaur, an immense reptile, associated with ammonites and other molluscan fossils in Cretaceous (Matawan) clay marls, at Merchantville, New Jersey. In *Annual Report of the State Geologist for 1896,* 248–250. Trenton: New Jersey Geological Survey.

Wyman, J. 1855. Notice of fossil bones from the red sandstone of the Connecticut River Valley. *American Journal of Science* 20:394–397.

Young, L. O., Jr. 1990. "Free-lance" searches for state's dinosaurs. *Baltimore Sun,* December 10, 1A, 2A.

————. 1991. Fossil hunters dig up biggest dinosaur bone yet in Northeast. *Baltimore Sun,* June 5, 1B, 4B.

I N D E X